Materials: History, Science and Perspectives

Louisette Priester

Materials: History, Science and Perspectives

Louisette Priester
Department of Materials Science
University of Paris-Saclay
Paris, France

ISBN 978-3-032-15753-9 ISBN 978-3-032-15754-6 (eBook)
https://doi.org/10.1007/978-3-032-15754-6

This Springer imprint is published by the registered company Springer Nature Switzerland AG
The registered company address is: Gewerbestrasse 11, 6330 Cham, Switzerland

Foreword

Traditional science education does not give much importance to "materials." From a very young age, the student will become acquainted with the physical states of "matter" using the example of water. A very complicated example, but justified by its omnipresence as well as the ease of experimenting and describing its phases in the ordinary conditions of our lives (this is also why water occupies the place it does on our planet!). Later, learning the periodic table of elements will allow us to become acquainted with simple bodies and their atomic properties. Later still, we will study some properties of these bodies and in particular their electrical resistance. But resistance is limited to a small drawing in the shape of a spiral, and not a real conductor in which a fraction of its electrons moves. Nothing wrong with this simplification: understanding electrical properties, in relation to the nature of the elements, is a matter of a course in solid-state physics as soon as we want to go beyond simplistic images.

It seems that, in all this progression, we have missed the point of "materials," and this for several reasons. First, because behind this term, some properties of use are drawn which require an integrated approach to the properties of matter, its shaping, the various elements which compose a material and contribute to its proper functioning. Furthermore, the study of materials requires knowledge of properties which are not even mentioned in classical teaching and which nevertheless condition what materials are. For example, we do not teach what the *elasticity* of a body precisely represents, for the good reason also that teachers have often not learned it during their training.

I remember the difficulty I had, in a meeting of a program committee of the 5th channel that was evaluating programs on materials, in making the members understand how incongruous the sentence that headed one of these programs could be "rubber is the only substance that has elasticity in its natural state..." It was doubly false: all solid bodies have elastic properties, as for rubber, it only becomes a solid once vulcanized. The *fluid mechanics* that we learned until recently in Faculty was only that of perfect fluids (which Richard Feynman in his famous physics courses called, somewhat derisively, *dry water* as opposed to *wet water* which has viscosity). The *friction* that one body exerts on another is especially not mentioned in high school: only frictionless contact that allows exercises in conservation laws in mechanics has a place in teaching. I had, very shortly before his death, the opportunity to interview Pierre Gilles de Gennes, who was very attached to the study of this theme; he began thus "the reflection on friction starts from a practical need." And it is here also that our teaching approach should be rethought.

Louisette Priester's book is an interesting contribution to this approach to promoting the theme of "materials" that will interest many curious people and training managers. Chapter 1 alone constitutes a small model work in a progression that goes from the early ages of humanity to today's materials, the archaeologist's material marking time like the paleontologist's fossil. Through these first stages of humanity, Homo Faber will gradually learn to master the exploitation and use of natural mineral resources. He will shape them, combine them. These will be the jewels, tools, and other weapons that are there to remind us of a history well before writing. In this sense, the teaching of materials is indeed interdisciplinary since in addition to the physicist, the chemist and the mechanic, the historian and the art teacher should be called upon.

The rest of his work follows a classic approach to the study of matter that can only be understood by agreeing to zoom in on the macroscopic properties of use down to the atomic scale without forgetting the intermediate scales, *a long journey in a large space* as L. Priester calls it. I like to talk about a 3M approach (macro, meso, micro). I was particularly interested in the third chapter which deals with *the life of materials*. If the thermal agitation of fluids is well understood, its manifestations in the case of solids are not as visible. We go through various manifestations: diffusion of atoms in a solid, reactivity and corrosion, changes in crystalline states under the effect of temperature or pressure, displacement of atoms under the effect of mechanical stresses. The aging of materials at the macroscopic level—*how long will it last?*—is a very active theme in research today and is influenced by these small-scale movements. This is the case of the very slow evolution of glasses and amorphous materials or the spontaneous settling of a column of stacked grains. A simple and spectacular experiment that I saw very recently at the ENS in Lyon can illustrate what is paradoxical in this evolution. A sheet of paper in which a transverse slit has been made is subjected to a traction force weak enough for the tear not to propagate. However, very slowly, the paper will lengthen in successive jumps following micro-tears. If we analyze this kinetics, we see that it corresponds to a thermal agitation effect at the level of the microfibrils that make up the paper. Who would have guessed? Following this part on the life of materials, a final chapter deals with properties of use and necessarily remains schematic. It is there to remind us that "materials" means applications of use.

Coming back to what I wrote above, I fear that I have been unfair in my somewhat peremptory and definitive judgment of primary and secondary education. On the one hand, the current effort around "*La main à la pâte*" aims at a better integration into primary school of an initiation based on observation and experimentation that relies on real situations or objects. There is clearly a place in the themes of investigation for the study of materials. Furthermore, at the middle school level, there is a technology teaching that starts from descriptions of complex material systems to arrive at their uses and properties. But this teaching, a distant heir to manual work, is too isolated from the general teaching of sciences and the classic disciplines of middle school.

Today, an experiment launched by the Academy of Sciences and the Academy of Technology is being set up, under the leadership of Pierre Léna and Yves Quéré, as an extension of "*La main à la pâte.*" It aims to combine science and technology teaching in an integrated way, in a single unit, by a single teacher. Around a hundred classes in France are currently participating. Quite naturally, the theme chosen for the first year of Collège is "matter and materials." It is too early to assess the experiment, even if there are initial indications from volunteer teachers and their students. I have the privilege of participating in this experiment in the preparation of a teacher's book that we have named "*Entrées en matières.*" This is not the implementation of a new program, but an effort to harmonize and coordinate the knowledge of various participating disciplines in a single Science and Technology project.

It is therefore quite natural that I agreed to read the documents prepared by L. Priester and to preface his work. I found there a rich source of information that will certainly be useful for this experiment and beyond. Because I am convinced that, in forms that will be defined according to the assessment of this experiment over 4 years, there will remain strong traces of an experiment that associates science and its applications. Madame Priester's book will be an element of this development, I am convinced.

Etienne Guyon

Preface

Materials have played an important role in the development of civilizations, from prehistory to the present day. Everything around us is material: *natural material* with wood, leaves, mud or stone houses… *material developed by man* with concrete buildings, glass or aluminum facades, clothing, means of transport, from the "*petite reine*" to the Airbus 380…. We cannot imagine a world without materials; man "lives with" by appreciating their performance certainly, but by taking little interest in their constitution and forgetting their necessity. The main objective of this book is to remedy this lack of knowledge, to interest the reader in this world of the "*non-living*" and however "*non-inanimate.*"

It is not a question here of taking up the notions of atom, ions, electrons, that is to say of the structure of matter, even if materials are described on these bases. Unlike the notion of matter, a notion of pure science which is sufficient in itself, that of material is inseparable from use. By material, we mean "*material for*" the manufacture of objects. *A material is therefore essentially useful*, which does not exclude approaching it scientifically. Knowing a material in order to control, or even master, its properties of use requires making a link between its physical, chemical, and mechanical characteristics. This is the object of *materials science* whose ultimate objective is to help in the choice of a material for a given use; this science has its originality, it is not reduced to any of its components: physics, chemistry, and mechanics.

The choice of a material is not only based on its functionality, it can also depend on *aesthetic and geopolitical criteria*. We must pay attention to the *availability* of its sources, its *durability*, its *cost* which unfortunately does not escape speculation. Increasingly, the absence of *pollution* risks and the possibilities of *recycling* are major assets for the candidacy of a material for the design of a new product.

A car, a turbojet engine includes 500–700 different materials. More generally, the choice of the appropriate material(s) for a given object is vast. However, despite their large number, basic knowledge of a limited number of them allows us to understand the characteristics of most of the others. This is why the materials cited here will be limited in number; to find out more, the reader can consult a list of works given at the end of this book. To gradually introduce the reader to the world of materials, we have divided this book into four areas. Chapter 1 discusses the *human, social, and historical aspects* by following the path through time of some typical materials, showing how they evolve with, for and thanks to man and trying to predict their future. The *scientific aspects* with the minimum basic knowledge on the constitution of materials, their differences, and their methods of development are briefly explained in Chap. 2. The *descriptions of the elementary processes* that take place in a material under the effect of external stresses and over time are the subject of Chap. 3. Finally, Chap. 4 presents certain uses of materials; far from being exhaustive, it attempts to show how various properties of materials are used in everyday life. The *ductility* of sheet steel allows it to be shaped for automobile bodies. The *semi-conductivity* of silicon is the basis for the design of the transistor and then of the electronic chip which led to the development of computers. The *superconductivity* of mixed oxides based on copper and barium leads to applications in energy storage and, in a more advanced field, allows the development of tomorrow's particle accelerators. Finally, the *giant magnetoresistance* of metal multilayers leads to the miniaturization of the reading heads used in computer hard drives.

New technologies are based on the improvement of traditional materials, on the development of new materials and, more recently, on the discovery that the properties of a material depend closely on its morphology. The response to a given stress of a material in a so-called "massive" form (three appreciable dimensions) and the response of this same material in the form of nanometric wires or films (one nanometer = one millionth of a millimeter) can present a large difference, even greater than that which exists between two different materials. Recent uses of materials are mostly based on their miniaturization: nanometric grains, thin layers, multilayers, etc. Whether natural or artificial, simple or sophisticated, with dimensions and shapes observable to the eye or requiring optical instruments, materials will always remain man's companions.

Paris, France Louisette Priester

Acknowledgments A first version of this book was published in 2008. The idea came to me following discussions I had with Sylvie Furois, a member of the Center for the Popularization of Knowledge (CVC) at Paris-Sud XI University. I tried to convince her of the value materials hold for a wide audience, and she encouraged me to demonstrate this. One of the main motivations for publishing this first book was my work with the French Society of Metallurgy and Materials (SF2M); I led a committee whose goal was to attract young people to scientific disciplines, including materials science, and which focused on training and employment in this field. I am particularly grateful to Jean-Marie Welter †, former president of the SF2M (2005–2006), who has always actively supported this effort. Several colleagues and friends helped me in writing the first edition of this book by providing information, documents, illustrations, and proofreading. I would like to thank them all, especially Jacques Castaing, Michèle Gupta, Olivier Hardouin Duparc, Jean Philibert, and Pierre Priester.

A few years later, the book was still in demand, but it no longer reflected developments in the world of materials. Drawing on a series of lectures I gave in various locations to a wide and varied audience, I revised the first part of the book to take into account the development of new materials, especially those that pose environmental and geopolitical problems. I would like to express my sincere gratitude to Etienne Guyon ‡, a researcher at the École Supérieure de Physique et de Chimie Industrielles de la Ville de Paris, honorary director of the École Normale Supérieure, and former director of the Palais de la Découverte, who agreed to write the preface to this book.

† Jean-Marie Welter (1944–2022).
‡ Etienne Guyon (1935–2023).

Competing Interests The author has no competing interests to declare that are relevant to the content of this manuscript.

Contents

1

Materials and Man

Men live with and thanks to an impressive quantity and diversity of materials whose origins, characteristics, and properties they are often unaware of. It is not a question here of making a catalogue, but of extracting some of them called "*classic*," such as wood, terracotta, iron, and aluminum, which have constituted and still constitute a large number of everyday objects. In addition, lesser-known materials, sometimes "*exotic*," are candidates for replacing some old materials or are grafted onto them to form an object. These recent materials do not, however, eliminate the use of traditional materials, they especially open the way to new uses in the technologies of the future and, as such, deserve our attention.

From Prehistory to the Present Day: What Materials?

Since prehistory, materials have always accompanied man and contributed to the development of civilizations. Throughout history, man has added artificial materials to natural materials, but they all fall under the same classification.

L. Priester, *Materials: History, Science and Perspectives*,
https://doi.org/10.1007/978-3-032-15754-6_1

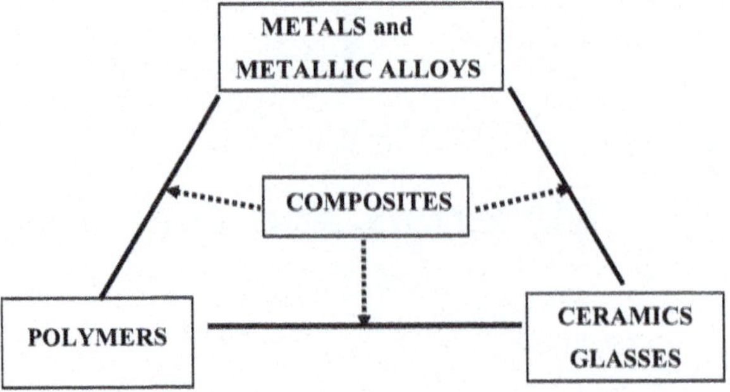

Usual classification of materials

(See Chap. 2 for a general definition of a material and each of the classes above)

Since prehistoric times, different classes of materials have been used:

- *Natural polymers* belong to the living, animal, and plant world: bones, cartilage, hair, wool, leather, wood, fabric, etc. Man has used them in a very diverse way: wood for bows and arrows, skin, bones and shells mixed with wood in statuary (Plate 1), fabric to protect against the cold (Plate 2). *Synthetic polymers,* which appeared in the twentieth century, are the basis of plastic materials that make up most of today's everyday objects.

Plate 1 Niki statuette (Democratic Republic of Congo): wood, fabric, skin, shell and feathers

Plate 2 Poncho, Caracas culture (600 to 300 BC)

- *Traditional ceramics and glasses* (Fig. 1.1) are mineral materials based on oxides* (mainly silicon oxide or silica for glasses)… The first tools and weapons are made of flint (ceramic) or *obsidian** (glass). The various terracotta pottery (produced from clay), necessary for daily life, also bears witness to the artistic tastes of our ancestors. The great architectural achievements, pyramids and temples, are made of stone (ceramic). *Technical ceramics*, including nitrides* and carbides in addition to oxides*, have been developed for less than a century. Always in the form of small pieces, they are used for their mechanical and/or thermal resistance, but the largest share of their market is in the electronics sector.
- *Metals and metal alloys* (obtained by adding metallic elements, or not, to the base metal) are relatively recent materials, although their first uses date back to around ten thousand years BCE. Between copper or gold, which is the first metal used by man? Both metals, very ductile, must have been used simultaneously in the manufacture of jewelry, but it is copper, exploited for useful purposes (hunting, war), which gives its name to the *archaeological period* called the "Copper Age" from 6000 to 3000 BCE, approximately. Nowadays, *metallic materials* are very widespread in the mechanical, energy, and transportation industries (about a hundred metal shades are found in an automobile engine).

Fig. 1.1 Lower Paleolithic flint cleaver and Roman amphorae

Archaelogical period	Dating (years before BCE)	Some important events (Relationship Man/ Materials)
Paleolithic (stone age)	2 million 1 million 450,000 180,000 70,000 40,000	- Oldest known tools (Africa) - *Homo habilis* separates from the Australopithecines - *Homo erectus*, first true representative of Hominids - *Homo erectus* lights the fire - *Homo sapiens* speaks - *Homo sapiens* including Cro-Magnon man *(modern man is related to this species)*
Mesolithic (middle stone age)	10,000	- Invention of the bow and arrow in Europe - Appearance of pottery in Japan

(continued)

(continued)

Archaelogical period	Dating (years before BCE)	Some important events (Relationship Man/ Materials)
Neolithic (age of polished stone)	9000 6000	- Appearance of the loom - Copper as a currency in the Mediterranean region
Copper age	5000 3200	- First sailing in Egypt - Invention of the wheel in Sumer - The first merchant ships sail the Mediterranean
Bronze age	3200 1400	- First written documents with pictograms (Sumer) - First bronze tools (Near East) - Invention of the plow - Construction of the pyramids (Egypt) - Use of bronze in China - Hittite Empire (Central Anatolia)
Iron age	1400	- Appearance of iron in the Near East then used in Europe - First network of major roads in Assyria - Invention of the wheelbarrow (China) - Invention of the paddle wheel (Near East) - Beginning of the Christian era

Overview of the different archaeological periods before our era and the main events affecting the relationship between man and the material. The age defined by archaeologists (Stone Age, Copper Age, Bronze Age, Iron Age) is associated with the main material used to make tools. However, this distinction is formal, because the real limits of the use of a material vary according to the cultural area and the geographical area.

– The term "*composite*," used for about half a century, refers to an intimate combination of materials generally belonging to different classes. However, composites made of various materials of the same class (wood, bone, paper) have existed for millennia. In a composite, the respective qualities of the constituents complement each other to form a material with improved overall performance.

From Empirical Knowledge to Materials Science

It was not until the seventeenth century that real knowledge of materials replaced the intuition that had prevailed since prehistory. Materials science is based on three disciplines: mechanics, physics, and chemistry; it cannot be reduced to any of them.

Some materials present in nature, such as wood or stone, have been directly shaped into tools. On the other hand, obtaining a fabric required know-how from the basic polymer, plant fiber. But it was above all metal, rare in its pure state, that proved recalcitrant before being worked by man. Initially extracted from colored rocks (ores), it was then hammered using a hammer on an anvil, both made of stone. Progress in the control of fire led to considerable progress in the exploitation of metal. Man had learned that fire warmed his cave and cooked his food, but it took millennia for him to discover that heated copper was more easily shaped. Then he carried out (around 4000 BCE) the fusion of ores, thus making it possible to obtain metals more easily and in larger quantities. Ores were often not available near the place where the metal was mined, and their necessary trade largely contributed to bringing peoples closer together.

Gradually and intuitively, man has been able to answer this question: what material for what use? Recent studies of a Gallo-Roman sword (Saint-Germain-en-Laye Museum, France) reveal a very clever design. The core of the blade is made of ductile (malleable) steel, while the periphery is made of a layer of hard steel: the blade can bend and, at the same time, withstand impacts with armor (Fig. 1.2). It is one of the first examples of "custom-made material," a concept of current technology and even more of the future.

But a true science of materials, with its mathematical language, only emerged in the seventeenth century. The first known law governs the deformation of materials, it is due to Robert Hooke (1635–1703). The crystalline arrangement is then described by René Just Haüy (1743–1822), finally Laurent de Lavoisier (1743–1794) brings the quantitative aspect to the chemistry of materials (Fig. 1.3). At the end of the eighteenth century, "Metallurgy" asserted itself with its three components: mechanics, physics, and chemistry.

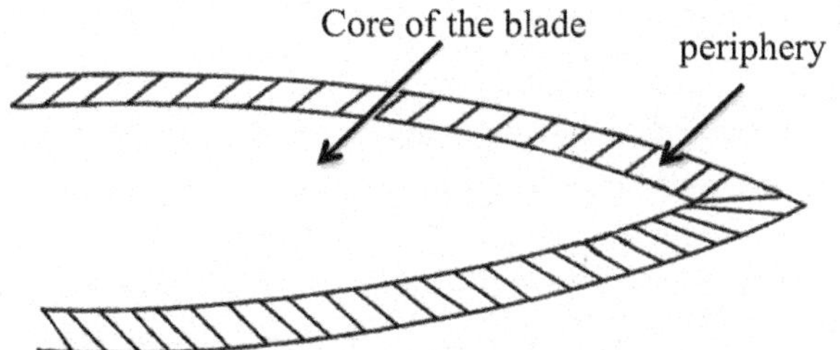

Fig. 1.2 The microstructure of the core of the blade makes it flexible, that of its periphery makes it shock-resistant

Fig. 1.3 Antoine Lavoisier and Marie-Anne Paulze, by Jacques Louis David, metropolitan museum of art

For two centuries, only metals were studied. The established concepts and developed approaches were then extended to other materials. Thus, we can say that Metallurgy is the paradigm of materials science.

From Gods to Man: Gold

Unalterable, considered to be of divine essence, gold produced at the nanometric scale is very reactive and humanized, however it loses none of its value.

Pure gold has a bright yellow color and is unalterable over time: these two characteristics led the ancients to attribute divine properties to it. Prehistoric gold objects have disappeared, probably melted down and the metal reused, hence the difficulty in deciding on the primary character of gold compared to copper or vice versa. From the mask of the pharaoh Tutankhamun (1343 BCE) made of 11 kg of solid gold (Fig. 1.4) to the great Buddha of Bangkok (fourteenth to eighteenth centuries) weighing more than 5 tons man's fascination with gold has spanned centuries and religions (Fig. 1.5). Gold is also the symbol of perfection the golden number*, the golden age, the sports medal awarded to the first winner of a competition.

Gold is a malleable metal that lends itself well to the creation of very fine wires (filigree) for jewelry. In goldsmithing, it is often alloyed with other

Fig. 1.4 Funerary mask of Pharaoh Tutankhamun, with which he was buried in 1343 BCE. The eyes and eyebrows are made of lapis lazuli

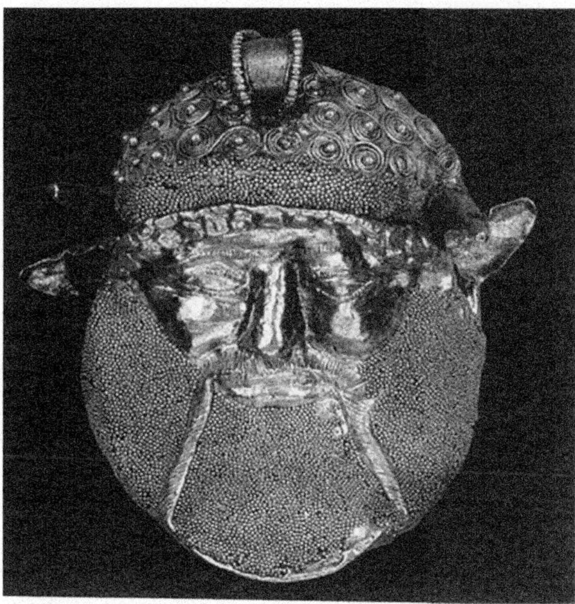

Fig. 1.5 Gold pendant in the shape of the head of Achellos—River God. Etruscan civilization (circ. 480 BCE)

metals to obtain various colors: yellow, pink, white, red, or green. Carats represent the mass percentage of gold in an object, 24 carats correspond to pure gold. Bankers and individuals buy gold in the form of ingots and coins, some to compensate for the issue of currencies, others to amass wealth.

The search for gold was one of the main motives for the great historical conquests. In the sixteenth century, the Spanish and Portuguese conquistadors plundered the gold of the pre-Columbian peoples in Mexico, Colombia. The discovery of these riches gave rise to a legend about the existence of a mythical land overflowing with gold: El Dorado (from the Spanish *el dorado:* the golden). Closer to us, in the nineteenth century, in the North American West, there was the *Klondike Gold Rush* (Fig. 1.6), recounted with great imagination in the cinema in Charlie Chaplin's film, in Blaise Cendras' novel and in comic strips (Scrooge McDuck, Lucky Luke). Even today, men risk their lives to extract gold nuggets using mercury, although it is known for its toxicity.

Nowadays, gold is not absent from high-tech industries, in the fields of electronics and space, where its inalterability is exploited. Almost all electronic components contain a small amount of gold: about 30 mg in a cell phone. Hundreds of millions of phones are produced each year, their average

Fig. 1.6 An endless nightmare: the climbing of the Chilkoot Pass by gold prospectors in the Klondike (late nineteenth century)

lifespan is about 2 years, and few of them are recycled at the end of their life: what a huge waste of gold for the economy!

Very recently, it was discovered that gold could lose its chemical inertness and acquire high reactivity if obtained in the form of *nanoparticles*, consisting of a few hundred atoms. It is then capable of trapping carbon monoxide, a deadly gas, transforming it into carbon dioxide, which is non-toxic to humans, even though it contributes to the greenhouse effect. If we can master the manufacture of gold nanoparticles, this metal, in its non-noble form, could lead to numerous applications in the automotive and chemical industries, in biology, and in medicine. This enters the field of *nanoscience*, where elements exhibit unique properties compared to those in their bulk form.

From Pendant to Chip: Copper

Copper is probably the first metal to be exploited: from the pendant of Shanidar (Irak—9500 BCE) to the connections between components of microelectronic chips, including electrically conductive wires.

The word copper comes from the Latin *Aes Cyprium*. Also called *cuprum* by Pliny the Elder in his Natural History (first century), it refers to the metal of Cyprus, a site renowned for the presence of this metal or its ore in Antiquity. But copper has been known for about 10,000 years by prehistoric men for whom it constitutes a new raw material compared to stone, wood, bone, etc.

Used since the Neolithic age for ornaments, as evidenced by the pendant found in Shanidar in Iraq around 9500 BCE (Fig. 1.7), copper then largely enters into the manufacture of coins, weapons, tools (Fig. 1.8), utensils, etc.

In the fifth millennium, active centers of copper extraction from smelted ore were found at Tal-I-Iblis (Iran) and Timna (present-day southern Israel). The importance of copper in the daily lives of people throughout the Near East is revealed by one of the earliest Sumerian texts that relates a dialogue between copper and silver, both personified:

> *"When winter comes, you do not provide man with the copper axe to cut wood. When harvest time comes, you do not provide man with the copper sickle to reap grain. Silver, if there were no palace, you would have no fixed abode... Like a god, you do not contribute to any useful work. How dare you confront me?"*
>
> *(Excerpt from a "debate" between copper and silver in person in which copper reproaches silver for its uselessness—Sumerian text from 2000 BCE)*

Pure copper or copper alloyed with zinc to form brass allowed the creation of the first scientific instruments: astrolabes*(Fig. 1.9), compasses, scales, and most of the metal parts used on board ships. Many wind musical instruments

Fig. 1.7 Malachite pendant (55% copper) exhumed From a cave in Shanidar (Iraq) around 9500 BCE

Fig. 1.8 Copper nail found at Tal-Iblis (Iran) around 4100 BCE

are also made of brass. Copper containers are used to make tasty dishes and to distill alcohols. The good thermal conductivity associated with the beautiful salmon pink color of copper makes it a metal still appreciated for pots and pans.

But the essential characteristic of copper is its very good electrical conductivity thanks to which it experienced a considerable boom at the beginning of the twentieth century, with the development of electricity. Its use is widespread in the distribution of electrical energy (95% of the conductive wires of an Airbus are made of copper) and in electrical and then electronic construction. In the field of microelectronics, copper provides the connections between the different components of chips and, with the increase in the performance of integrated circuits, it also serves as a basic conductor inside the chip.

A Legendary Alloy: Bronze

The first bronze objects, discovered around the Dead Sea, date back to the third millennium BCE. Bronze appeared later in China, but the artisans of the Shang Dynasty (1600–1100 BCE) remain the undisputed masters of working with this alloy.

The Bronze Age spans from about 3000 to 1400 BCE. Bronze is an alloy of copper and tin, the tin content varying from about 3 to 25% depending

Fig. 1.9 Copper was used for early scientific instruments such as brass astrolabes (an alloy of copper and 30% zinc), made here by the French craftsman Jean Fusoris around 1400

on its use. At the very beginning, it was an alloy of copper and arsenic from a natural combination of the two metals in ores, numerous throughout the Near East. Did the death of some blacksmiths, caused by the release of toxic arsenic gas during the smelting of the ore, play a role in the replacement of arsenic with tin? Why tin? Was it after trying different additions that the smelters chose tin, which gives a harder and less brittle alloy than arsenic bronze? The mystery remains.

Bronze was the first alloy used by man to make tools and weapons, more robust than these same objects in copper, then common containers of various shapes, musical instruments, and various sculptures (Plate 3). An impressive number of bronze objects were found in the royal cemetery of Ur where the Sumerian kings were buried with their entire retinue from the third millennium BC. Such a discovery, astonishing because tin was rare in the Near East, suggests the existence of trade between this part of the world and the heart of Europe where tin ores were found.

Plate 3 Bronze carved deer, Alacah Hüyük, circa 2400 BCE

The bronze workers of the Shang Dynasty (1600–1100 BCE) made containers, utensils, sculptures, weapons, and armors whose meticulous castings demonstrate a know-how superior to that of Westerners (Fig. 1.10). This is all the more surprising since the Chinese seem to have gone directly from the Neolithic to the Bronze Age. Was their use of bronze an influence from the Near East or a rediscovery? There is no way of saying this. In reality, only the aristocracy used bronze tools and weapons during hunting parties or war expeditions, while the peasantry continued to live in the Stone Age.

The fabulous bronze krater (large vase where wine and water were mixed) discovered at Vix, the site of a Gallic oppidum, in a sixth-century BC tomb, is one of the first European bronzes (Plate 4).

Plate 4 Art Nouveau lamp "The Barbary Fig Tree", bronze mount, Majorelle, 1902

Fig. 1.10 Bronze wine vase with four legs decorated with clouds, dragon eyes, and various abstract motifs. The handle of the lid is shaped like a small elephant like the vase itself. (Shang Dynasty, 1600–1100 BCE)

During the Renaissance, bronze was the material of excellence for sculptors: one of the doors of the Baptistery of Florence by Lorenzo Ghiberti (completed in 1452) made Michelangelo say that it was worthy of being the door to paradise (Fig. 1.11).

More recently, Art Nouveau, Art Deco, and then contemporary art did not forget bronze (Plate 5). Apart from sculpture, bronze rich in tin (20 to 25%) is used to make bells. Bronze remains a heavy and expensive alloy, which certainly limits its wider use.

Plate 5 Bronze krater discovered in the tomb of a Celtic princess at Vix (Côte-d'Or), circa 530 BC Musée du Pays Châtillonnais, Châtillon-sur-Seine

Fig. 1.11 The apparition of the three angels and the sacrifice of Isaac, one of the Bas-reliefs of the "gate of paradise" of the baptistery of Florence (gilded bronze, 1425–1452) by Lorenzo Ghiberti

From the Hittite Dagger to the Millau Viaduct: Iron and Steel

Iron metallurgy: an important step in man's progress toward history. despite its meteorite origin, iron quickly became a metal within everyone's reach.

Although iron was known as early as 4000 BCE, the Iron Age did not begin until around 1500 BCE. Why so late? Two reasons: iron fusion, requiring a high temperature (above 1600 °C) was impossible in the furnaces of the time and this metal rarely existed in its natural state. It could then be extracted from meteorites that fell from the sky (Fig. 1.12).

A pin fashioned from native iron and dating back to 3000 BCE was found in a Hittite site (present-day Anatolia) at Alacah Hüyük.

In the first millennium BCE, iron metallurgy developed in Europe, already accompanied by pollution! Iron sulfide*, roasted by the Minoans to obtain the metal, polluted the air, causing complaints from Cretan farmers. The use of iron spread into the Celtic world: the hooping of wooden chariot wheels with iron constituted a revolution in transport as important as the later use of rubber tires. Today, iron products are everywhere in metal constructions, workshops, houses, gardens, and fences.

Fig. 1.12 Meteorite iron and goethite, an ore composed of iron hydroxide and red iron oxide (hematite)

A major handicap of iron is its sensitivity to corrosion in most surrounding environments, leading to rust. Thus, seaside dwellers are frequently called upon to sand and cover their fences with red lead anti-rust paint. However, the famous iron pillar of the Qutub Minar in New Delhi (fourth century) has suffered almost no corrosion damage. This resistance fuels debates among metallurgists. Could it be the phosphorus present in low-level iron that, by concentrating on the surface of the object, forms a protective layer with the atmosphere? Is this then a deliberate addition or an unpredictable consequence of the method of obtaining iron? The question remains.

Between iron and its main alloy, *steel*, there are a few millennia, but above all a low percentage of carbon (less than 1.8%) which leads to large differences in properties. Various metals (nickel, chromium, titanium, etc.) are added to the binary alloy to obtain *alloyed steels* with specific properties, including *stainless steels*. Steels are widely used in transport and construction. Car bodies are mainly made of steel containing very few additives. Stainless steel is widely used in major architectural projects where it contributes to the strength and elegance of structures (Fig. 1.13).

If the carbon percentage in iron exceeds 1.8%, cast iron is obtained, long dedicated to the manufacture of household utensils. Around the eighteenth

Fig. 1.13 The Millau Viaduct is a reinforced concrete and steel structure 2,460 m long and measuring up to 343 m high. 36,000 tons of steel were needed for its construction (four times the weight of the Eiffel Tower)

century, the development of cast iron frames for the construction of commercial and industrial buildings was very successful in the United States: the buildings were lighter and less expensive than those made of stone. Cast iron was also used in the construction of facades in place of marble or granite: balustrades and cornices imitated fashionable architectural motifs (Fig. 1.14).

Fig. 1.14 Building facades with cast iron columns and cornices in the Soho District of New York (early nineteenth century). Once painted, cast iron looks deceptively like stone

From the Imperial Eagle to the Can: Aluminum

Although it has lost its titles of nobility, aluminum continues to gain notoriety in many industrial sectors and in the world of architecture.

Aluminum is a recent metal compared to copper and iron, it dates from the beginning of the nineteenth century. Discovered in 1827 by Friedrich Wöhler, it was shown at the Universal Exhibition in Paris in 1855. At that time, the exploitation of its ore, bauxite (name coming from its discovery near Baux-de-Provence, France) was very expensive. Aluminum was considered a noble metal, intended for the manufacture of jewelry, semi-precious objects (jewelry boxes, cigarette cases, etc.) and the reproduction of antique statues. This first use of aluminum is based on two of its remarkable qualities: a very low density and a high resistance to oxidation in air, its appearance remains unchanged over time. Thus, in 1860, Napoleon III ordered 17 imperial eagles in gilded aluminum for flagpoles, weighing three times less than the same ones in bronze (Fig. 1.15).

Fig. 1.15 Aluminum imperial eagle for the flagpole of a cavalry regiment under Napoleon III

Until the end of the nineteenth century, the market share of aluminum was low because its exploitation, although improved by Henri Sainte-Claire Deville in 1846, still did not allow mass use. In 1887, Karl Joseph Bayer developed a less expensive method of obtaining it, aluminum then entered the industrial world and, at the same time, lost its value for the jeweler.

Did Jules Verne have, in 1865, a premonition of the role that aluminum would later play in aeronautics, by imagining in *From the Earth to the Moon* the conquest of space with an "all-aluminum cannonball"? On Earth, in 1899, we discovered the "Jamais contente," an electric car made of aluminum alloy (partinium) which reached a speed of 100 km/h (Fig. 1.16).

During the First World War, aluminum became a strategic metal with the Breguet 14 bombers, the structure of which was mainly made of duralumin (an aluminum alloy with 4% copper). It also temporarily found its use again for curious jewelry, the rings that the poilus made with the debris of German rockets: "*In aluminum, pale as absence and tender as memories*" according to Guillaume Apollinaire. The hood of the Citroën DS 19 presented at the motor show in 1955 is also made of aluminum (Fig. 1.17).

Fig. 1.16 The "Jamais contente," an electric car made of aluminum reached a speed of 100 km/h in 1899

Fig. 1.17 The Citroën DS 19 has an aluminum hood

More exotic is the use of aluminum for the making of a dress by Paco Rabanne (1966) (Fig. 1.18).

Since the middle of the twentieth century, the major industrial markets for aluminum have been packaging (household paper, beverage cans, etc.), construction (metal frames, etc.), architecture (building facades), and aerospace, where it is generally used in the form of alloys. The wheels of the interplanetary robot launched on Mars by NASA in 2004 are made of aluminum, because they must be light and, at the same time, strong and flexible. Anodizing* aluminum gives it a variety of colors and increases its resistance to corrosion.

Aluminum presents a major asset for the future: it is almost infinitely recyclable, without loss of quality, and this operation is less expensive than its manufacture from bauxite.

Fig. 1.18 Aluminum dress by Paco Rabanne (1966)

Other Metals and Alloys: Tradition and Innovation

A large number of metallic materials are part of tradition, but new metals and alloys are emerging in the industrial world for increasingly targeted applications.

The range of metallic materials, although less extensive than that of polymers, nevertheless covers a great diversity. While classic metals such as iron and copper were widely used at the beginning of our era, many of them only emerged in the eighteenth century, such as chrome, cobalt, zinc, or zirconium, and even in the twentieth century for titanium. All have now become *traditional*. More recently, and for very targeted applications, *exotic* metals have been introduced into industry: lithium, beryllium, palladium, etc.

Among *traditional metals*, *zinc*, known for its resistance to atmospheric corrosion, is the metal of gutters, building roofs, and galvanized coatings* of steels (47% of its consumption). Car bodies are made of galvanized steel. *Superalloys*, based on *nickel* and *cobalt*, have good mechanical strength and good resistance to hot corrosion. With aluminum and titanium alloys, they are widely used in aeronautics (Fig. 1.19).

Titanium, the production of which was not really developed until 1939, appears in all industrial sectors. Resistant to corrosion and fire, it equips the Ariane 5 engine and constitutes the armor of aircraft carriers as well as the hull of several Russian nuclear submarines. In the automobile industry, a titanium spring is four times lighter and takes up half the space of its steel equivalent. Its biocompatible nature makes it a material of choice for prostheses

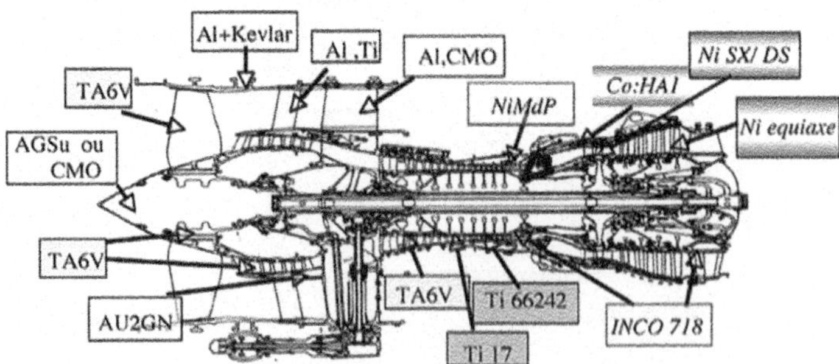

Fig. 1.19 Materials in a turbojet: we note the importance of alloys based on aluminum (white box), titanium (gray box) as well as nickel and cobalt (italic acronyms). The alloys are heated to different temperatures: aluminum base and TA6V up to 200 °C, Ti17, Ti 66,242, NiMdP and INCO718 up to 650 °C; those based on cobalt and nickel (degraded box) can reach 1,700 °C because they are mechanically resistant and resistant to the oxidizing atmosphere of the gases emitted by turbojets

Fig. 1.20 Anodized titanium facade of the Guggenheim Museum in Bilbao (Spain) (1997) (Architect Frank Gehry)

and dental implants. Anodized, it shines in all colors and is weather-resistant, hence its use in architecture (Fig. 1.20).

By combining 50% titanium and 50% nickel, we obtain a so-called *shape-memory alloy* whose surprising property is to recover its initial shape after being significantly deformed, provided it is heated slightly. Thus, crumpled eyeglass frames adopt their complex shape again. Nickel, titanium, zirconium, and magnesium form *intermetallic compounds* (see Chap. 2) that are promising for storing hydrogen, a non-polluting energy vector.

Exotic metals find applications in the IT and new energy sectors. Lithium, the lightest of metals, constitutes the batteries that equip laptops, digital devices, etc. Palladium, in the form of nanowires, has the remarkable property of detecting hydrogen, a promising characteristic for the automotive industry of tomorrow. The exotic aspect here falls under the domain of nanoscience, and it is associated with the dimensions in which the metal is used and not with its nature.

Rare Metals: Geopolitical Issue

A rare metal is defined by its very low percentage in nature. This notion of rarity has evolved a lot over time. Aluminum, considered rare before 1900, is today one of the most used metals. We can also consider as rare: gold, silver, and some metals used in special steels such as molybdenum, niobium, tantalum, actinide elements, and all radioactive (from thorium to lawrencium). The platinoids (platinum, palladium, iridium, rhodium, ruthenium, and osmium) are distinguished especially by their rarity: three of them (Pt, Pd, Rh) are used as catalysts for catalytic converters for automobiles, and iridium and ruthenium are used in the electrical and electronics industry.

Rhodium is the rarest metal on Earth (the rarest element is a metalloid: astatine), but it is also the most expensive metal in the world, it is now 15 times more coveted than gold.

Rare Earths: The Black Gold of the Twenty-First Century

The lanthanide series, chemical elements with atomic numbers between 57 and 71, to which scandium and yttrium are grafted, forms the rare earth group. Contrary to what their name suggests, these earths can be quite widespread in the earth's crust, like some common metals. These are 17 metals which, thanks to their remarkable optic and electromagnetic properties, have conquered high-tech markets: electronics, television, hybrid vehicles, missiles, etc., not without posing geopolitical problems since China is the leading producer, representing nearly 61% of annual global production (in 2021). The majority of the remaining mining production is provided by the United States, Myanmar, Australia, and Thailand.

Electric and hybrid cars (Fig. 1.21), electric two-wheelers, offshore wind turbines, loudspeakers, mobile phones, household appliances, and military technologies need magnets to work. Dysprosium-doped neodymium-iron-boron magnets, currently the most efficient, are the reason for a large part of rare earth mining.

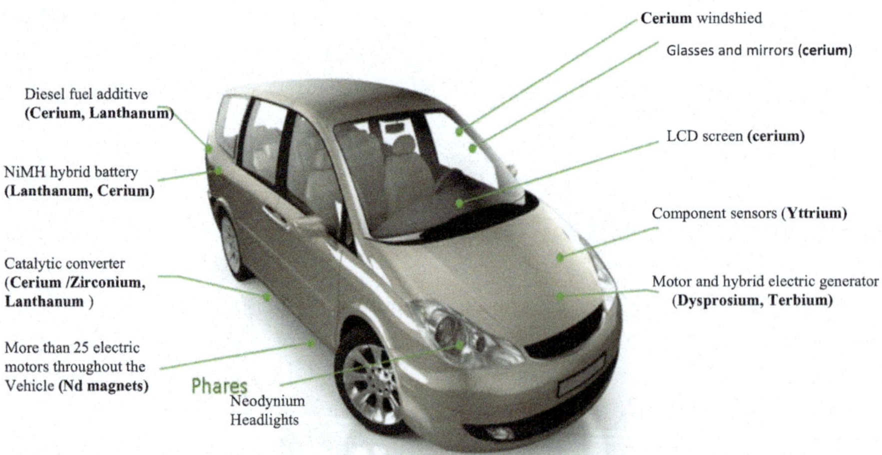

Fig. 1.21 Electric and hybrid cars can contain 9 to 11 kg of rare earths

These needs for rare earths will continue to increase in the coming decades: replacement of gasoline-powered vehicles, defense industries, microelectronics, renewable energies (photovoltaic and wind) and, more innovative and competitive, the development of artificial intelligence.

A fundamental contradiction lies in the fact that the evolution toward a greener world is dependent of dirty metals whose odious mining conditions lead to high rates of cancer and respiratory diseases among the workers who extract them. A large majority of the extraction sites are located in Inner Mongolia, an autonomous region of China, which has become a veritable hell on Earth (Fig. 1.22).

The major nations, unable to accept this quasi-monopoly of China, are trying to compensate for their lack of resources by researching the recycling of rare earths. The difficulty lies in the fact that the latter are often present in very small quantities and that it is difficult to separate them from other metals in order to recycle them. Hopes are emerging with, among other things, the possibility of using molecular sponges, porous materials based on metal ions and organic molecules, which have extraordinary absorption capacities, capable of recovering traces of rare earths in order to recycle them. It also seems possible to replace lanthanum with magnesium. As is often the case, difficulties, in this case, the supply of rare earths, are at the origin of increasingly strategic innovations.

Fig. 1.22 Lake that has become toxic due to the extraction of rare earths, near the city of Baotou, the largest industrial city in Inner Mongolia. Photo Euronews

From the Amphora to the Rotor: Ceramics

Fragility is the Achilles heel of ceramics. Despite this handicap, ceramics allowed the transport and preservation of food in Antiquity thanks to amphorae. They are now used in rotors and bulletproof vests.

Natural ceramics are major constituents of our planet: *rocks* (sandstone, limestone, granite, etc.) are the oldest building materials and *ice* is of prime importance in preserving the environment.

The word ceramic comes from the ancient Greek *keramos*, which means "potter's earth" and which gave its name to an Athens *Keramikos* District. The same term is used to designate the material and the objects made from this material. *Traditional ceramics* or *terracotta* are first used to store and transport food: jars and amphorae. They are made from clays, the most common being *kaolin**, mixed with other minerals to give them a particular color. Their use remains important in the current fields of tiles and tableware.

Artistic ceramics appeared in the Far East around the 10th millennium BCE, well before Greek pottery. It takes on a particular perfection in China, a country which currently remains the leading exporter of ceramics in the world (Fig. 1.23).

The art of pottery has endured through the centuries and even found a resurgence of activity at the end of the twentieth century with the introduction of techniques from Japan (raku*). In France, many sites maintain their tradition: Gien, Moutiers, Longwy (Plate 6).

Fig. 1.23 Celadon ram from the Eastern Jin Dynasty, China (317–420)

Plate 6 "Colonial ball", Art Nouveau ceramic vase, Longwy enamels, created by Maurice Paul Chevallier for the Paris Colonial Exhibition, 1931

From the twentieth century onward, *industrial or technical* ceramics were added to utility ceramics and art ceramics. At the top of technical ceramics, we find the hardest of all the natural materials: diamond! When it has a defect (internal bubbles or foreign particles), the diamond loses its value for the jeweler, and is then used in cutting and drilling tools or as an abrasive. *Structural ceramics*, with interesting thermomechanical properties, are used to make engine parts in automobiles, tap seals, etc. Rigid, wear- and shock-resistant, they constitute the ballistic plates inserted into the tactical bulletproof vests of police special intervention forces, they provide effective protection for vital organs against assault rifle ammunition. Among *biomedical ceramics*, alumina, which has no reaction with the physiological environment, is used in the manufacture of dental implants and hip prostheses (Fig. 1.24).

However, *technical ceramics* have mainly established themselves in the field of *electronics* due to the diversity of their electrical and thermal properties. They represent 90% of the industrial ceramics market: electrical insulation, capacitors, missile guides and, more recently, substrates and interconnection modules in microelectronics. In particular, silicon, thanks to its semiconductor properties, has enabled the creation of integrated circuits (chips) and photovoltaic solar cells*.

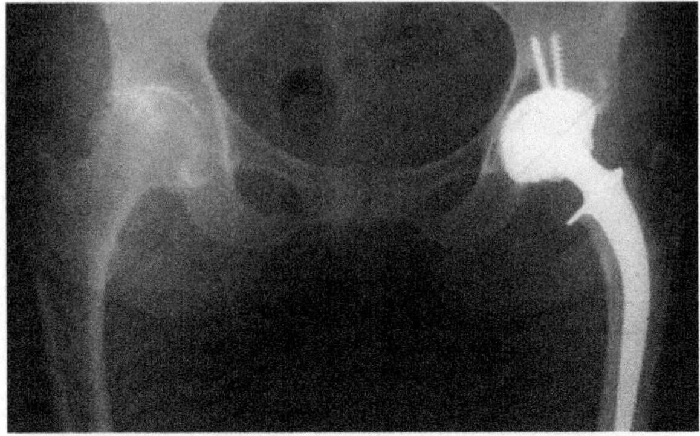

Fig. 1.24 Hip prosthesis is composed of two interlocking parts: one replaces the articular part of the pelvis (acetabulum) and the other replaces the neck and head of the femur. Alumina, used for both parts, has the advantage of resisting wear and preventing the destruction of bone tissue

Finally, *cement* (a mixture of lime, silica, and alumina) is also considered a ceramic. It is used on a gigantic scale in the manufacture of concrete, which replaces stone in large-scale constructions.

From Obsidian to Solar Energy Panels: Glass

Obsidian*, resulting from the cooling of volcanic lava, is a natural glass. Early humans used it to make small, sharp tools obtained by breaking the material. This brittleness must be controlled in the industrial glass used to make large solar panels.

The natural process of volcanic glass formation contains all the basic ingredients for making industrial glass: sand, metal oxides, and heat. Small glass vessels have been used since the 3rd millennium BCE in the Middle East and Egypt. They were then *molded* into clay and sand cores. Around 250 BCE, the Babylonians invented the *glassblowing rod*, allowing for the rapid production of larger vessels (Fig. 1.25). Babylonian glassblowers were invited to Rome, and, according to Pliny the Elder, glass cups gradually replaced gold cups among wealthy Romans.

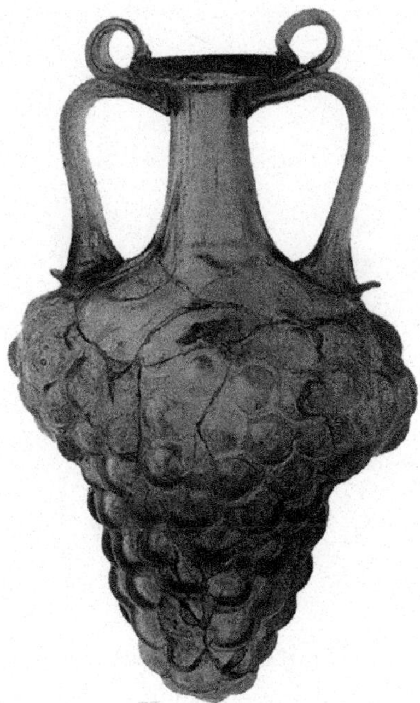

Fig. 1.25 Grape-shaped bottle; mold-blown glass—Rhenish production? Second–third centuries

Glass then suffered a long eclipse in Europe, apart from the creation of stained-glass windows in Gothic cathedrals (Plate 7). It reappeared in the *Quattrocento* in Venice, and more particularly on the island of Murano, where glassmakers adopted Eastern expertise. The *glassblowing technique* continues today, but machines have replaced humans to produce enormous quantities of bottles, light bulbs, telephone line insulators, etc. As for large panels of flat glass used in windows, they only really appeared in the nineteenth century, leading to a significant change in housing. The facade of an office building can have 80% glass surface. Glass also contributes to the elegance of new architectures (Figs. 1.26 and 1.27).

Plate 7 Stained glass window from the portal of Chartres Cathedral, 12th century. Detail from the life of Christ: the Annunciation to the Magi by the Star

However, humans are not the largest producers of glass. Diatoms (Plate 8), single-celled algae living in fresh or marine water, protect themselves with a glass shell whose highly varied shapes demonstrate the omnipresence of symmetry in nature (Fig. 1.28). The mass of glass they synthesize is significantly greater than human production.

Fig. 1.26 The Louvre pyramid, formed of diamonds and triangles in clear glass mounted on a steel frame (1988) (architect I. M. Pei)

Fig. 1.27 The glass building of the Louis Vuitton Foundation (2014)—(architect Frank Ghery)—resembles a sailboat with sails billowing in the wind, giving the illusion of movement. Photo Daniel Rodet

Plate 8 Marine diatoms seen under an optical microscope. The glass shells that protect them have surprising shapes

In practice, glass is used primarily for its optical properties: lenses, windows. It also constitutes most containers in the chemical and food industries because it reacts very little with most products. Absorbing some radiation, it helps contain nuclear waste. Glass wool is an effective insulator in buildings. Car windshields are made of laminated or laminated glass, consisting of a sandwich of a layer of plastic between two layers of glass.

Glass is far from absent from contemporary art. Glass paste obtained by firing colored glass powders and granules in a refractory plaster mold is used to create jewelry, vases, sculptures, and more.

Fig. 1.28 Plate by Erns Haeckel depicting diatoms, 1904. The glass shells that protect them have shapes dominated by symmetry

From Wooden Wheels to Formula 1 Tires: Polymers

Polymers are tending to replace other materials in many applications. Eyeglass lenses are no longer made of glass but of organic material. A polymer is replacing the enameled cast iron in bathtubs. We are witnessing a transformation of materials.

Natural polymers are the first materials of humanity. Humans learned to assemble wool or cotton threads and thus create fabrics to clothe themselves or wrap their dead in shrouds, thus giving one of the first signs of civilization. They worked plant fibers to obtain leaves for writing on papyrus and then on paper. They shaped wood for defense and then used it to build their houses, boats, chariots, and more. *Wood* is a remarkable material. It is a composite made of various polymers whose structure prefigures that of today's fiber-glass laminates. After a period of decline, it has once again become a material of choice for housing (Fig. 1.29 a, b). When cut lengthwise, its mechanical strength can be very high; it is also an excellent thermal insulator, its efficiency being 15 times greater than that of concrete and 430 times greater than that of steel. Finally, it has aesthetic qualities: depending on the tree and its cut, wood presents different colors and different appearances. It is a warm and environmentally friendly material.

Synthetic polymers include plastics that are either thermoset (reversibly hardened) or thermoplastic (reversibly malleable), and elastomers. The *plastic materials* that appeared at the end of the nineteenth century remained based on natural materials: *celluloid,* still used in ping-pong balls, and then *galalith,* highly prized for making jewelry (Fig. 1.30). The first truly synthetic polymer, discovered in 1907, was *Bakelite,* now sought after as a collector's item. *Polyester* followed in 1941, which was mainly used to make synthetic textile fibers, particularly in sports, and then *Formica,* an emblem of the modernity of the postwar boom. All three materials are *thermosetting.*

Fig. 1.29 From the mountain chalet benefiting from the thermal insulation qualities of wood (**a**) to the contemporary bioclimatic house (architect Damien Gallet) that respects energy efficiency, the environment, and sustainable development (**b**)

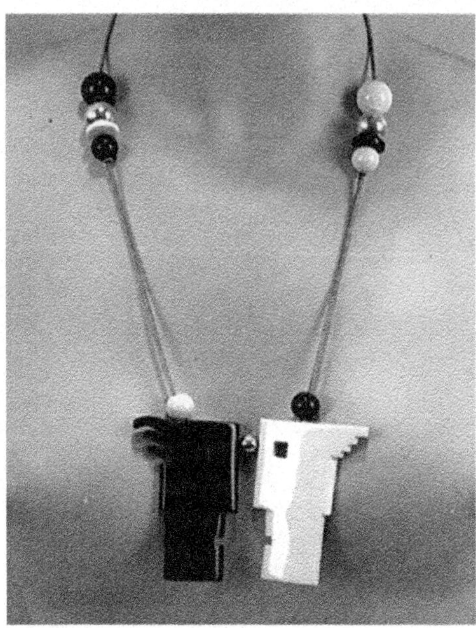

Fig. 1.30 Bauhaus Galalith Necklace by Guillemette L'Hoir (Bauhaus is an architecture and applied arts movement founded in 1919 by W. Gropius, which played a major role in the development of modern ideas and techniques)

Among the countless *thermoplastic polymers* that appeared in the first half of the twentieth century, *PVC* (polyvinyl chloride) and *ABS* (acrylonitrile butadiene styrene) are particularly notable. *PVC*, discovered in 1926 and industrialized in 1931, is primarily used for the manufacture of profiles and tubes, but also for floor and wall coverings, tarpaulins, electrical cables, roofing membranes, and more. Since the 1970s, *ABS* has gradually replaced Formica as a thermal insulator, but it is primarily the material of choice for bathrooms and household appliances. A famous toy manufacturer uses it to make its building bricks, and Citroën uses it as the body material for its Mehari model. *Kevlar* arrived at the same time as *ABS*. It is a synthetic fiber, not classified as a plastic material, but very strong, lightweight, and rot-proof. *Kevlar* boasts excellent tensile strength, at least five times stronger than steel of equal weight. It replaced asbestos due to its fire resistance (oven gloves, firefighter helmets), and was used in sports equipment (skating, skiing, snowshoeing, automotive, competition, etc.). Finally, it is used in the form of tightly woven fibers in bulletproof vests.

Elastomers (natural rubber and rubbers) are polymers with astonishing elastic properties. Car tires have long been made of rubber; having become more complex, they now use other polymers and metals.

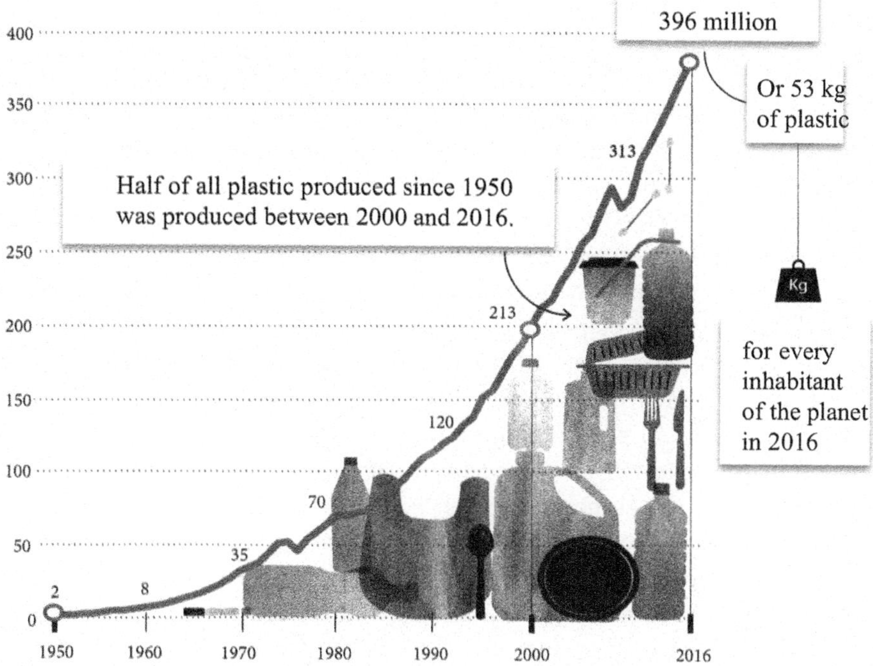

Fig. 1.31 Global plastic production, in millions of tons per year (2016), mainly for packaging (40%)

From the mid-twentieth century, there was a significant diversification and an explosion in demand for synthetic polymers (Fig. 1.31). Global production increased from 1.5 million tons in 1950 to 117 million in 1999, reaching 460 million tons/year in 2020. We are surrounded by plastics, in the form of consumer goods, in every sector. Who hasn't bought something made of polyethylene, polyester, epoxy resin, or nylon? There are more than 4,000 varieties of polymers, each with its own formula and manufacturing process, which is kept secret by its manufacturers. The field of *plastics processing*, encompassing all the processes and techniques for producing plastics, is considerably broader than that of *metallurgy*.

Farewell to Plastics

The twentieth century was the era of plastics. The craze for this material was expressed by Roland Barthes in his Mythologies (1957), where the author refers to it as "Miraculous Matter" or "Alchemical Substance." But at the end of the century, humans became aware of the danger of plastic pollution and are trying to reduce its production and, at the same time, ensure its recycling, a strategic issue for the economy. As waste, plastics accumulate in the oceans, forming what is called the seventh continent (an area equivalent to six times that of France), consisting mainly of packaging. Every minute, the equivalent of a garbage truck full of plastic waste is dumped into the oceans.

Plastics are a source of significant pollution even before becoming waste. Their danger affects every stage of their life cycle, from the extraction of fossil fuels needed for their manufacture to their disposal. They pose a global threat to the environment, health, and the climate (Fig. 1.32). Their recycling remains at 10% of production, while 80% of plastic products end up as waste in less than a year. Unfortunately, two major powers are opposing the reduction of their production: United States and China, respectively, the largest consumer and producer of plastics.

Arguing that the negative perception of plastics should not overshadow their advantages, two solutions are currently being considered to address their low recycling rate. The first option would be *bio-sourced polymers* made from biomass waste (wheat, corn, sugarcane, etc.), but this would require intensive farming, which is contrary to ecology! The second solution would use *biodegradable plastics* that degrade in the presence of microorganisms, but this would be difficult to implement. Unfortunately, forecasts for the coming decades still indicate an increase in plastic production, a curse from which it seems impossible to escape.

Fig. 1.32 Albatross found dead on a beach in the Midway Islands (North Pacific Ocean), its stomach full of plastic, September 2009.

From Natural to Industrial Composites

Harnessing the advantages of one material while minimizing the disadvantages of the other is the challenge of composites, inspired by the observation of certain natural polymers.

Wood and bone are *natural composites*. Bone is composed of a soft substance in the center and a hard substance around the edges. Wood is formed from *heartwood* (perfect wood), the main mass of the trunk, composed of dead, lignified cells that support the tree, and sapwood (imperfect wood), formed of concentric layers of cells that have not yet lignified, which ensures the conduction of raw sap (Fig. 1.33). Paper and concrete are *composites manufactured* long before the industrial era.

Currently, the term "*industrial*" is reserved for *composites* made of materials belonging to different classes: *CMP, CMM,* and *CMC* stand for Polymer, Metal, and Ceramic Matrix Composites, respectively. By combining a lightweight polymer with low stiffness and average mechanical properties with a rigid but brittle glass, a *CMP composite* material is obtained, whose properties differ from those of the basic components. It is used in a large number of components in the aeronautics, construction, defense, medical, rail, nautical, and urban furniture sectors, among others. It is well known as a material of choice in sports and leisure: skis, tennis rackets, and more. Reinforcements in *ceramic matrix composites (CMCs)* can be glass or carbon fibers, but also ceramic. This type of composite is used in space vehicles (heat shields and steering fins), disk brakes, and rolling bearings. Reinforced concrete is

Fig. 1.33 Cross section of wood showing light sapwood and dark heartwood

one of the first examples of *CMCs* in which the reinforcement constitutes a true metal frame. In *CMM* composites, inclusions in the form of small ceramic spheres or fibers (alumina, silicon carbide, or tungsten, etc.) are introduced into the metal matrix to achieve properties intermediate between the ductility of the metal and the hardness of the ceramic. *Cermets* (a neologism composed of the abbreviations *ceramic* and *metal*) were among the earliest *CMMs*. Composed of a dispersion of small ceramic particles (often carbides) in a metal matrix, they are commonly used to manufacture cutting tools and as armor for tanks. More recently, they protect space probes by resisting high-speed impacts from micrometeorites and orbital debris.

Generally speaking, the use of composites instead of metal alloys in transportation (trains, trucks, automobiles) has resulted in significant weight savings. However, despite the breakthrough of composite materials, metals maintain their supremacy in the transportation sector.

Materials of the Future: Toward Miniaturization—Nanomaterials

A nanomaterial has at least one external dimension at the nanoscale (between 1 and 100 nm); the nanometer (one millionth of a millimeter) is a value difficult to visualize in everyday life; it corresponds roughly to 1/50,000 of the thickness of a human hair. Another definition is based on the notion of characteristic length associated with a property. Indeed, any physical property is characterized by a length below which it no longer manifests itself; at the nanoscale, most materials exhibit exotic properties. We have already mentioned the extreme reactivity of gold in nanoscale form, in contrast to its inertness when it is in bulk.

There are two types of nanomaterials: *nano-objects* (Fig. 1.34) and *nanostructured materials*. Depending on the number of their nanometric dimensions, *nano-objects or nanoparticles* (3D), *nanofibers* or *nanotubes* (2D), *nanosheets* or *nanoplatelets* (1D). *Nanostructured materials* are made of nano-objects.

Nanoparticles are found inside cosmetic products (nail polish, makeup, etc.) or in certain foods (infant milk, confectionery, cereal bars, pastries, frozen desserts, etc.). Among these nanoparticles, the most commonly used are silver nanoparticles and *fullerenes* (molecules formed from 60 carbon atoms arranged around a sphere, the total size reaching approximately 1 nm). Ferromagnetic iron oxide nanoparticles, loaded with a drug that acts on tumor cells, can be guided from the outside onto the cancerous target by a magnet.

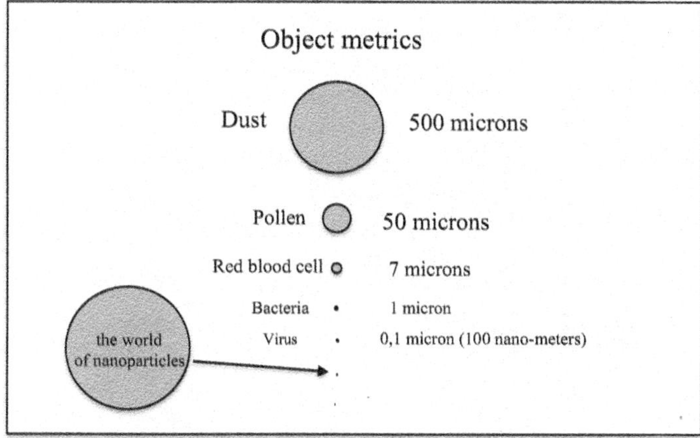

Fig. 1.34 The field of nanoparticles

First observed in 1991, *carbon nanotubes* appear as hollow, concentric tubes arranged around 60 carbon atoms (Fig. 1.35); sometimes there is only a single tube with an internal diameter on the order of a nanometer. To realize the infinitely small of this structure, one need only to know that a carbon nanotube with a length equivalent to the distance between the Earth and the Moon (384,400 km), rolled up on itself, would occupy the volume of an orange seed. Its length can be on the order of a few micrometers. Such a filament has a strength 100 times greater than steel, while weighing six times less. The electrical, mechanical, and thermal properties of carbon nanotubes suggest numerous applications, particularly in microelectronics and materials for hydrogen storage. However, industrial applications are slow to emerge, although they continue to fuel the wildest fantasies: space elevators, metal-free bridges, etc.

Are synthetic nanoparticles a danger to humans and the environment? This question raises many concerns. Given their size, they can infiltrate everywhere and travel to the brain via the nerves, or cause inflammation via the blood; the two combined effects can lead to cardiovascular effects. They therefore have a certain toxicological potential. However, these effects mainly affect professionals involved in the production of nanoparticles who must work under drastic protective conditions.

Among nanostructured materials, previously well-known oxide dispersions have gained resistance to irradiation and high temperatures with the introduction of nanoscale dispersed particles, giving the steel all the qualities needed for use in the reactors of the future.

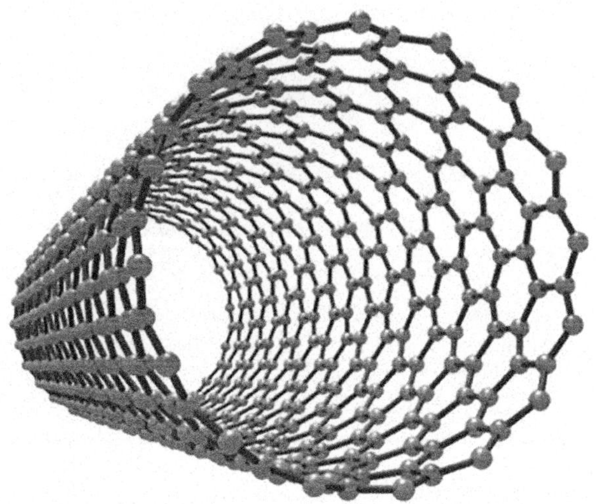

Fig. 1.35 Carbon nanotube, with a total thickness of 2 nm, each node representing a carbon atom. The length can reach from 100 to 1 nm

All the composites mentioned above can be reproduced at the nanoscale. For example, a nanocomposite made of clay nanoparticles (a ceramic) in a polymer matrix, used to coat tennis balls, can modify the inflation properties: air no longer escapes from the core of the ball.

Finally, *nanoporous materials* (nano-sponge, nanofoam, etc.) can be attached to nanostructures, with the nanoscale pores constituting the nanoparticles. Solid/gas composites are inspired by nature and lead us toward this other trend in materials of the future.

Materials of the Future: Nature's Lesson—Biomimetic Materials

In its efforts to develop new, more efficient materials, humans are turning to nature, whose natural materials often exhibit remarkable characteristics. In particular, many materials in the living world are bio-composites with fabulous properties: this is the case, among others, of eggshells and *mother-of-pearl.*

We have seen that ceramics are fragile and break upon impact. However, many biological materials have a ceramic structure and are impact resistant. The *mother-of-pearl* of marine shells, composed of layers of aragonite (Co3Ca + traces of Pb, Zn, Sr) alternating with layers of polymers, is a good example. Inspired by abalone mother-of-pearl, a new ceramic material, a true artificial mother-of-pearl, has been developed. It consists of 500-nm-thick aragonite platelets, which any potential cracks must pass through, thus delaying their propagation (Fig. 1.36).

Silicon carbide foam (Fig. 1.37) combines the hardness, high-temperature durability, and exceptional performance of solid silicon carbide with the ultra-lightness and versatility of foam. The combination of these properties makes it a highly sought-after product in many industrial sectors, such as aerospace, defense, and semiconductor manufacturing.

Miniaturization and biomimicry combine to create magnificent material structures analogous to those of living organisms, down to the nanoscale. Nanotubes, nanowires, nanofibers, and nanopowders are now found in all cutting-edge technologies: electronic nanocomponents, nanosensors, etc.

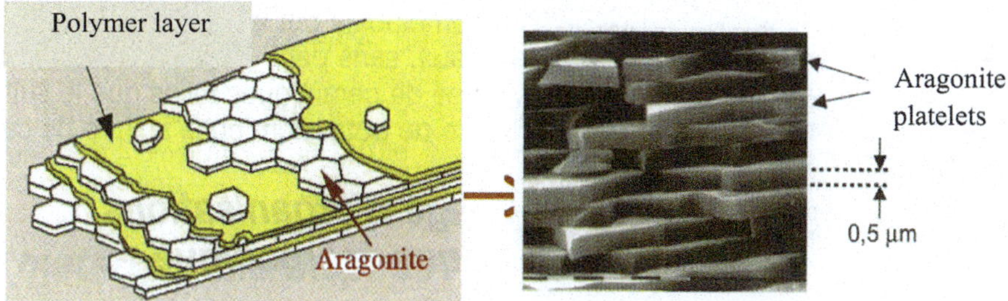

Fig. 1.36 Microstructures of a natural (left) and an artificial right "mother-of-pearl"

Fig. 1.37 Silicon carbide foam (ceramic) used as a support for metal catalysts to clean car exhaust gases

2

Discovering the World of Materials

In this chapter, we aim to familiarize the reader with the basic concepts of *materials science*. First, we define what we mean by a *material*, without focusing on describing the elementary atomic constituents. Then, we introduce what essentially differentiates materials: the *notion of order*. There are ordered materials and disordered materials, although the distinction is not as formal. A globally ordered material can be, like everything else in nature, *perfect or imperfect*. Defects here are not necessarily harmful; on the contrary, they give life to materials, most of whose functional properties are governed by their defects. The organization of atoms and defects in a material is expressed by its *microstructure*, a notion which, beyond the prefix "micro," covers different scales. Moreover, a material is rarely pure; foreign elements are distributed homogeneously in the base material, like wine in water, or in a non-homogeneous manner, like oil in water. Domains with different physical properties are thus formed: their existence is expressed by the concept of *phase*. The description of the microstructure integrates the number, the extension, and the distribution of phases. Knowledge of the *relationship between the microstructure and the properties* of a material is the basis for any use of it. This is the fundamental objective of materials science.

What is a Material?

A material is a matter for a material. This concise definition clearly distinguishes matter, of a physicochemical nature, from material, inseparable from its use to obtain an object.

L. Priester, *Materials: History, Science and Perspectives*,
https://doi.org/10.1007/978-3-032-15754-6_2

The terms "matter" and "material" lead to two methods of classifying materials. A first classification (Page 2, Chap. 1) is based on the nature of the bonds between atoms (covalent, ionic, or metallic bond) which are the basic building blocks of matter (Figs. 2.1, 2.2, and 2.3). Depending on whether the atoms remain isolated or group together to form molecules, the properties of the material differ considerably.

Metals and metal alloys are characterized by the absence of molecules; the atoms preserve their individuality. The *metallic bond* (Fig. 2.3) is ensured by

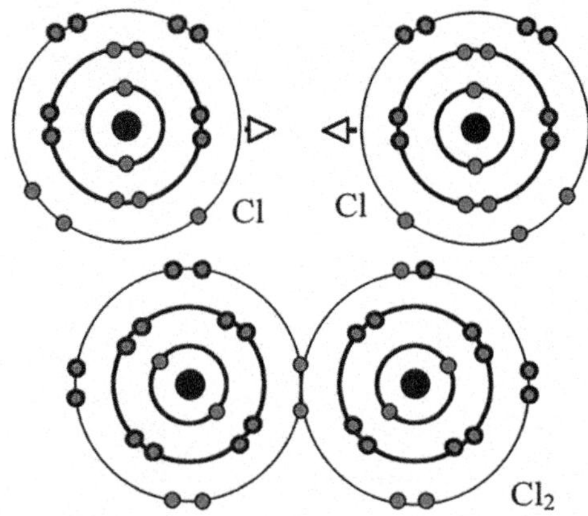

Fig. 2.1 Covalent bond ensured by the sharing of two electrons in the outer shell. ●: nucleus

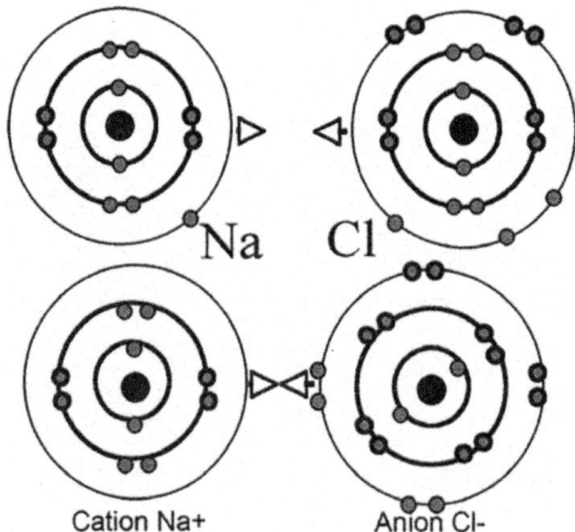

Fig. 2.2 Ionic bond ensured by the transfer of an electron from one atom to another. •: electrons (electrons orbit the nucleus—valence electrons are located in the outer shell)

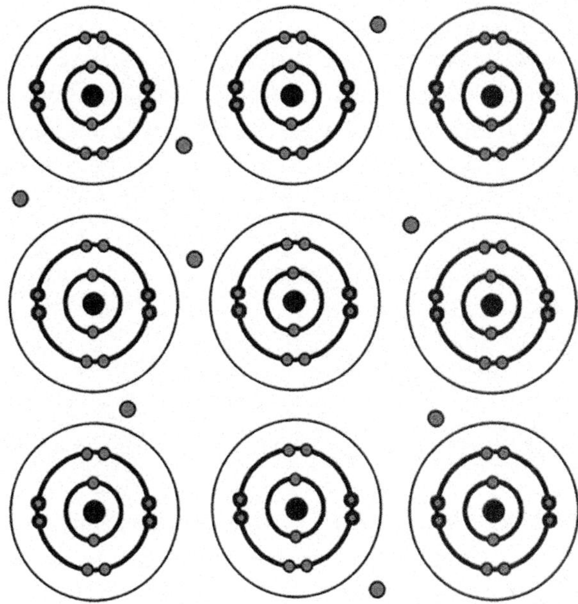

Fig. 2.3 Metallic bond provided by a "gas" consisting of free and delocalized valence electrons (•) that move throughout the solid. The movement of negatively charged electrons in one direction causes a positive current in the other direction

valence electrons free to move throughout the material, which explains the electrical conductivity of metals.

Polymers are *organic materials* in which atoms form *long molecular chains or macromolecules* whose backbone is made of carbon atoms. Two types of bonds ensure the stability of the structure: *covalent bonds* between carbon groups, additional covalent bridges, or *weak bonds** between macromolecules.

There is no definitive definition of ceramics and glasses; these two terms are commonly used to group together diverse materials that are neither metallic nor organic. The nature of the chemical bonds can help differentiate them: primarily covalent in glasses, covalent or ionic, often mixed (ionocovalent), and more rarely metallic in ceramics. But it is the technology that clearly distinguishes ceramics from glass, depending on the order in which the products are manufactured:

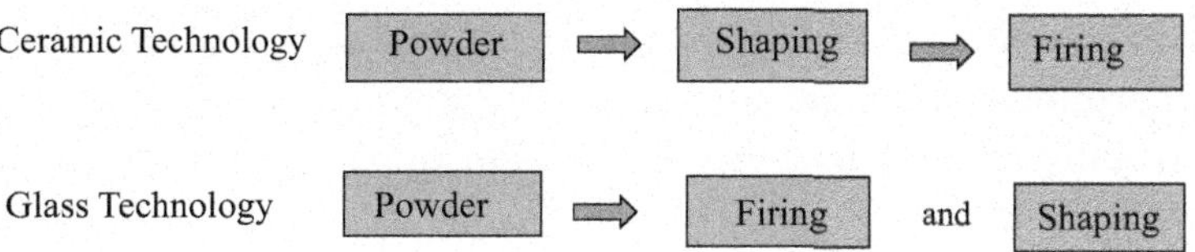

A second classification related to materials distinguishes between *structural and functional materials*. The former includes high-tonnage materials (steels, wood, cement, etc.) used primarily for their mechanical performance and resistance to various aggressive environments. Functional materials are chosen for their physical properties: optical (missile glass), thermal (ceramic resistors), magnetic (magnets), and electrical (semi- and superconductors). However, any part made from these materials must be formed and must withstand various stresses and environments over the course of its lifetime. Therefore, the structural properties of a material cannot be neglected, even when targeting a so-called "functional" use.

Order and Disorder in Materials

Order and disorder distinguish crystalline materials from amorphous materials, the archetypes of which are, respectively, metals and glasses. However, both states can coexist: local disorder within order, ordered regions within disorder.

A fundamental distinction between materials, beyond their classification into metal, polymer, and ceramic, is based on the notions of order and disorder, which have always had ethical connotations: order is traditionally associated with "Beauty." Disorder can, however, be a generator of wealth. By hammering metal, which creates a large number of defects (disorder) in its structure, humans have made it more resistant. The association between order and beauty is called into question in the field of materials. The crystal service so precious to us is made of a disordered material (glass): this crystal does not contain crystals!

Different levels of order are defined. *Long-range order* (Fig. 2.4) corresponds to the strictly periodic repetition, by translation in the three directions of space, of a pattern of atoms (Fig. 2.5). The pattern itself exhibits *short-range order* (typically less than 10 times the atomic distance). The combination of short-range and long-range orders defines crystallinity.

The crystal model with atoms in the form of spheres, each attached to a site, is a *rigid model* far from reality. The respective dimensions of the atoms and the cell are not respected. Furthermore, since the atoms are in perpetual *thermal agitation*, their average positions are periodically arranged.

In amorphous structures, the atoms are distributed randomly over large distances, but there is a local organization of the atoms relative to each other (Fig. 2.6).

Fig. 2.4 Ceiling of the dining room of Casa El Capricho near Santander (Spain-1883/85—Architect A. Gaudi.) This photo provides an example, on a macroscopic scale, of a two-dimensional translation network. The pattern contained within the square is translated in two directions by a distance equal to the side of the square (network mesh) to reproduce the whole

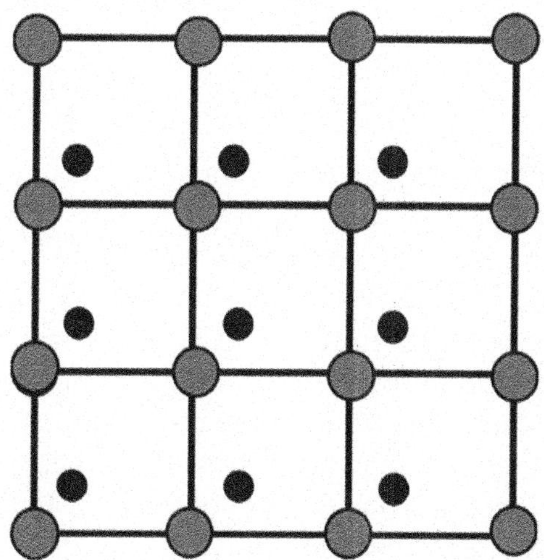

Fig. 2.5 Translation lattice: the basis vectors form a cell whose vertices are the nodes of the lattice. The pattern is a set of atoms plotted at each node of the lattice. It is characterized by the nature of each atom and its site in the cell, here two atoms: one atom (gray circle) at the node and one atom (black circle) located at a quarter of the diagonal of the square

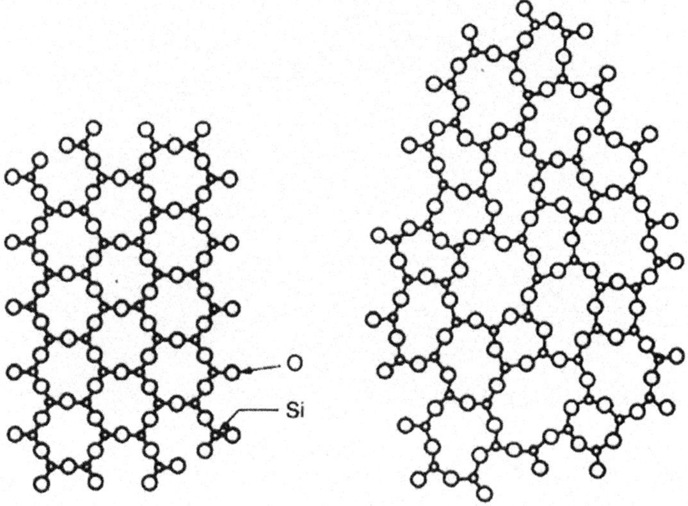

Fig. 2.6 Two-dimensional projection of a silica crystal. In space, each silicon atom is surrounded by four oxygen atoms forming a regular tetrahedron* (left). The tetrahedra persist in amorphous silica, but are distorted and, above all, form an anarchic stack (right)

An amorphous material therefore exhibits a certain short-range order. Similarly, a crystalline material always contains local imperfections or disorders. However, the separation between the two states remains clear and, like any dichotomy, constitutes an approximation. A *state of intermediate order* (at medium distance) may exist. A degree of order is then defined, and the associated structures are semi-crystalline. This is the case for many polymers: the crystalline regions result from regular stacks of macromolecules, and not atoms as in metals, and are surrounded by an amorphous tangle.

Unlike crystalline materials, which exhibit a classical architecture, that of polymers evokes the strange constructions of the Catalan architect Gaudi (late nineteenth–early twentieth centuries).

The distinction between materials in terms of their crystalline, non-crystalline, and partially crystalline structure (Fig. 2.7) does not cover their classification into metal, ceramic, and polymer.

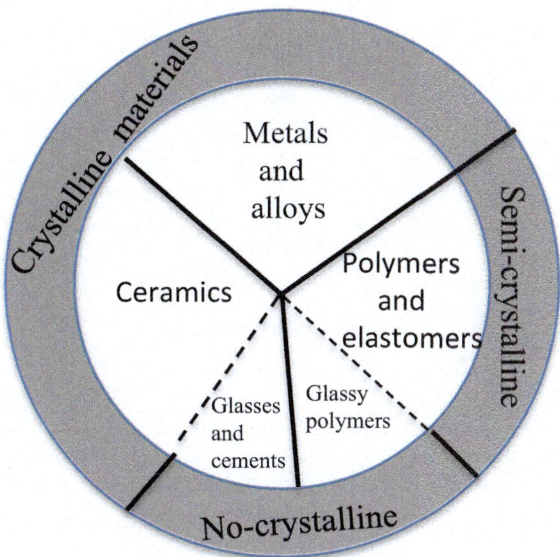

Fig. 2.7 Presentation of the different materials according to their structure: crystalline, non-crystalline (amorphous), or semi-crystalline

The Crystalline State—Perfect Order

A crystal is the elementary volume of a crystallized material. Nature and industry rarely provide a material in the form of a single crystal, but more often in the form of a group of crystals called a polycrystal.

A crystal is a periodic arrangement of atoms obtained by translation, along a lattice, of a *unit cell*. The nodes of the lattice are distributed on *planes* and *rows*. The parallel planes form a *family*. The different families of planes (and rows), which intersect in the crystal space, have varying degrees of importance for the crystal properties. Planes of high atomic density are distinctive; their cohesion is stronger and the distance between them (called inter-reticular) greater the denser they are (Fig. 2.8). They most often constitute the external faces of natural crystals. This is the case for quartz crystal (crystalline silica) that appears in the form of a hexagonal prism, with each of its six flat faces parallel to a dense plane of its structure (Fig. 2.9).

The arrangement of crystals in a polycrystals may be modelized (Fig. 2.10).

Metals most often adopt a cubic structure, with one atom at the center of the cube (body-centered cubic) or with one atom at the center of each face (face-centered cubic), or a hexagonal structure (Fig. 2.11).

Stackings are often more complex in crystalline ceramics, as the unit cells can contain a large number of atoms (Fig. 2.12).

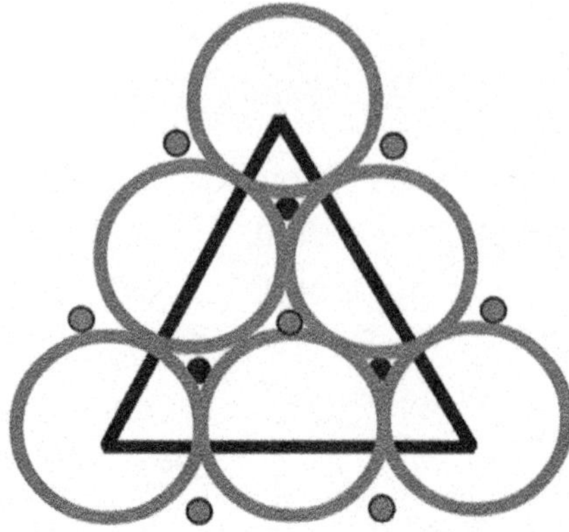

Fig. 2.8 Close-packed atoms in a dense plane. The large open circles represent atoms in the plane of the figure. The small gray and black circles represent the projections of the atom positions in the dense planes, respectively, above and below the plane of the figure

Fig. 2.9 Quartz crystals: six-sided prisms; each one ends in a pyramid

Fig. 2.10 Modeling: (left) a single crystal based on a Kelvin polyhedron; (right) a polycrystal constructed with these polyhedra. The light-colored figures correspond to the base and lateral layers. In a real polycrystal, the different crystals are not similar

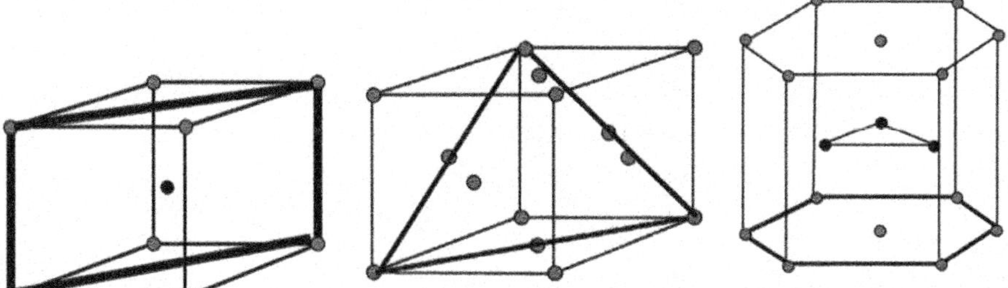

Fig. 2.11 The most common crystal structures (from left to right): body-centered cubic, face-centered cubic, and hexagonal. The atoms represented by black circles are inside the unit cell. For each structure, the smallest distance between atoms is located in the dense plane (delineated by thick lines)

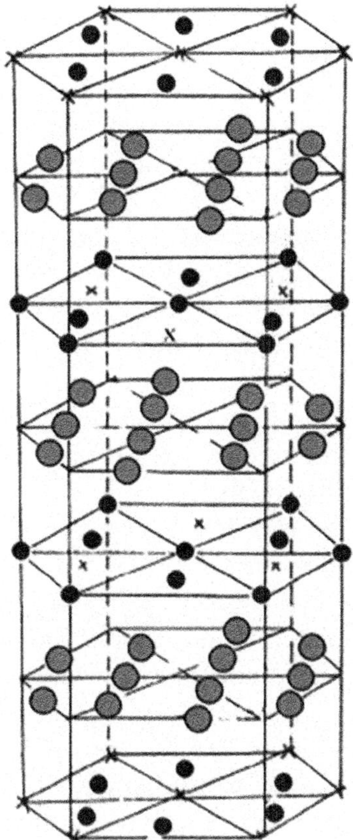

Fig. 2.12 Schematic representation of a half-cell of alumina (aluminum oxide). The total cell has twice the height. ●—Oxygen atom, ◎—Aluminum atom, x—Aluminum hole

There are necessarily voids in these stacks; the highest compactness corresponds to only 74% of the total volume. The interstices allow the incorporation of small foreign atoms into the structure. This results in distortions of the crystal lattice, but above all, this dissolution of an element in the crystal is the origin of the formation of so-called interstitial alloys.

Crystalline Imperfections

Defects govern many of the functional properties of crystalline materials. They can be detrimental by causing corrosion and breakage, but contrary to their name, they are often beneficial to the crystal. They allow metal to be shaped or hardened, and glass to be colored.

Crystals are theoretically perfect only at an absolute temperature of 0 K, i.e., 273 °C below 0 °C, a temperature never reached in practice. Defects are always present, the nature and distribution of which depend on the *thermomechanical history* of the material, i.e., all the heating, cooling, and deformation it has undergone during its development, shaping, and subsequent use. Crystalline defects are of four types, depending on their dimensional order: 0D, 1D, 2D, or 3D.

Point defects (0D) are *vacancies* (Fig. 2.13), *interstitial* (Fig. 2.14), and *substitutional atoms* (Fig. 2.15). All move under the effect of heat, giving rise to the phenomenon of *diffusion*, one of the most important in the "life" of the material. The dimensions of the atoms (apart from hydrogen) that fit into the structure of a crystal are always larger than those of the possible insertion sites. Interstitials therefore occupy only a fraction of the sites, as long as the deformations they cause are tolerated by the crystal. Similarly, steric reasons limit the size of the substitutional element in the crystal lattice.

A dislocation is a linear (1D) defect. The core of the dislocation (a region extremely close to its line) is surrounded by a distorted region of the crystal (Fig. 2.16).

Dislocations, as they move, are the basis for the (permanent) plastic deformation of a solid, which is one of the characteristics of crystalline materials (see Chap. 3).

Around the middle of the twentieth century, disturbances in the crystal lattice were visible using electron microscopy (Fig. 2.17b and c), but the presence and distribution of dislocations had been indirectly revealed on the surface of a crystal by corrosion patterns. Indeed, the distorted regions around dislocations are attractive to foreign atoms, which feel cramped in the perfect regions of the crystal. Impurities decorate the dislocations, which are then cleverly detected through selective corrosion of the surface (Fig. 2.17a).

In a polycrystal, the boundaries between crystals (also called grains) are planar (2D) defects or grain boundaries. The orientation of the crystal lattice differs on either side of a grain boundary (Fig. 2.18).

Volumetric (3D) defects include pores, inclusions, and precipitates. A pore is a cavity within a crystal or at the boundary between grains (Fig. 2.19). Inclusions and precipitates are particles: some are of a different nature from the material; others appear through reactions between the material's constituents. The precipitation phenomenon is used to harden a material by varying the size and distribution of the particles (Fig. 2.20).

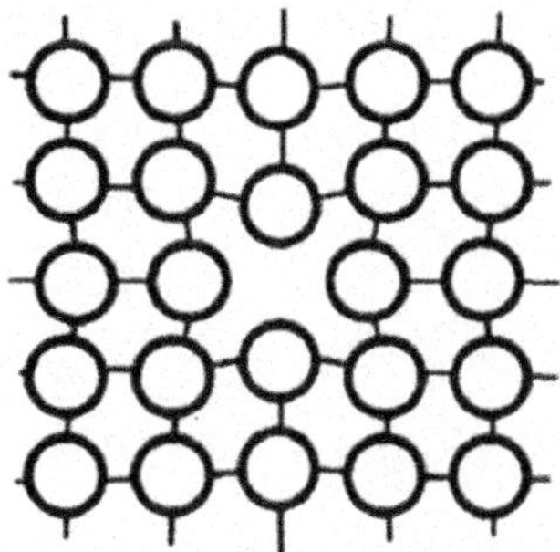

Fig. 2.13 A vacancy is an empty site) in the normal location of an atom

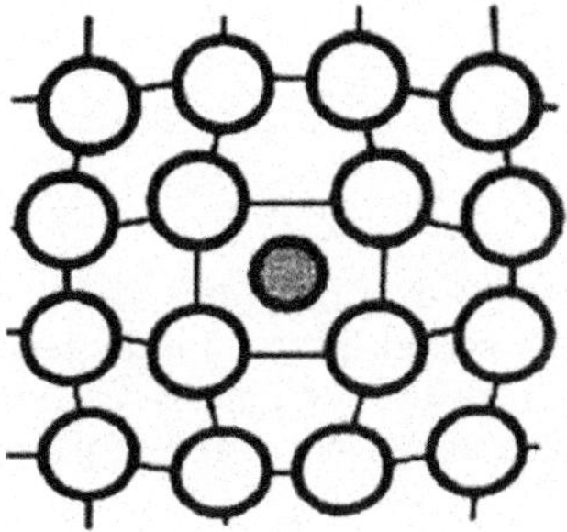

Fig. 2.14 An interstitial atom (gray in a crystal lattice (white)

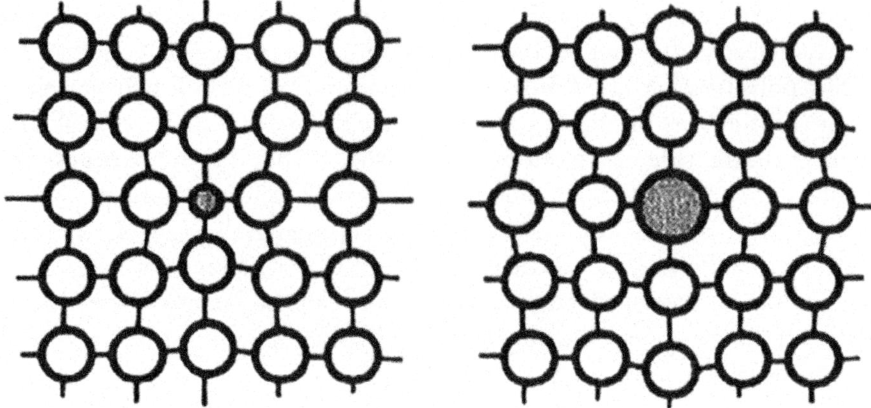

Fig. 2.15 A substitutional atom (gray): smaller or larger than the normal atom in the crystal lattice

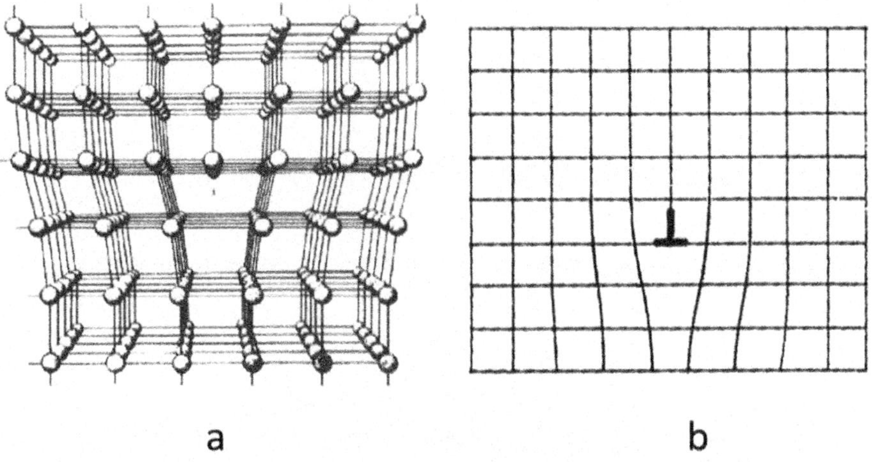

Fig. 2.16 Three- and two-dimensional diagrams of a dislocation: An additional plane of atoms (vertical plane at the center of the image) causes a curvature of the neighboring crystal planes (**a**). The dislocation line is the edge of this plane (⊥), but in reality, the defect concerns the entire disturbed region surrounding the line (**b**)

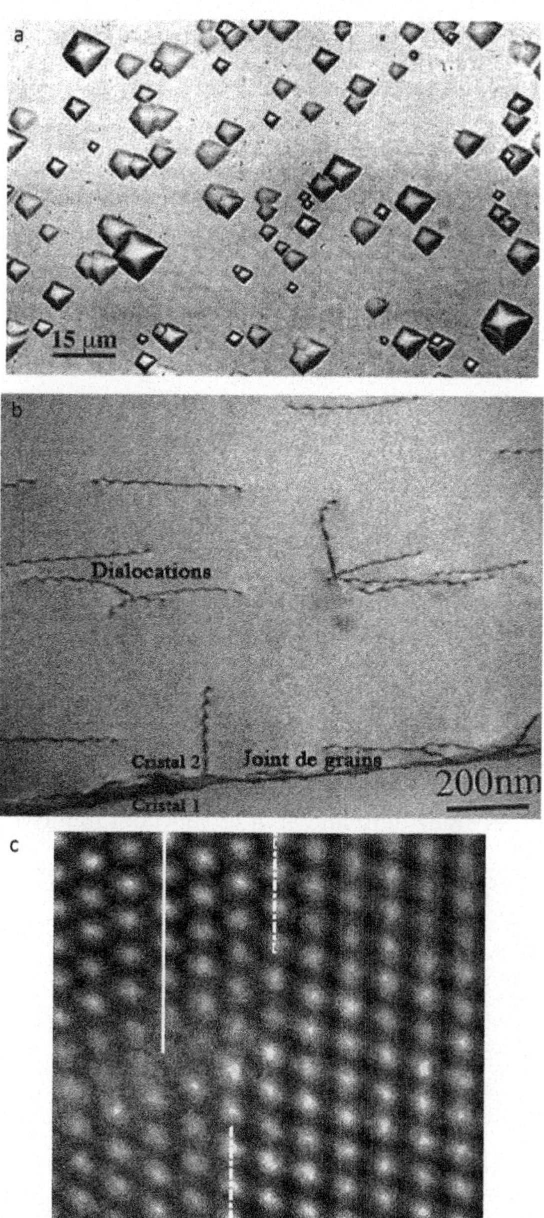

Fig. 2.17 **a** Dislocation distribution in a nickel crystal revealed by corrosion patterns. **b** In transmission electron microscopy, dislocations appear as contrasting black lines (the oscillating appearance is due to interference of the electron wave with the material). The diagonal line at the bottom of the figure separates two crystals: this is a grain boundary. **c** High-resolution electron microscopy (atoms in white) clearly shows the additional vertical plane (solid line) and the distortions of neighboring planes. The dislocation (edge of the additional plane) is near a grain boundary (dotted lines); the change in orientation (from the left crystal to the right crystal) of the atomic planes relatively inclined with respect to the vertical is clearly visible

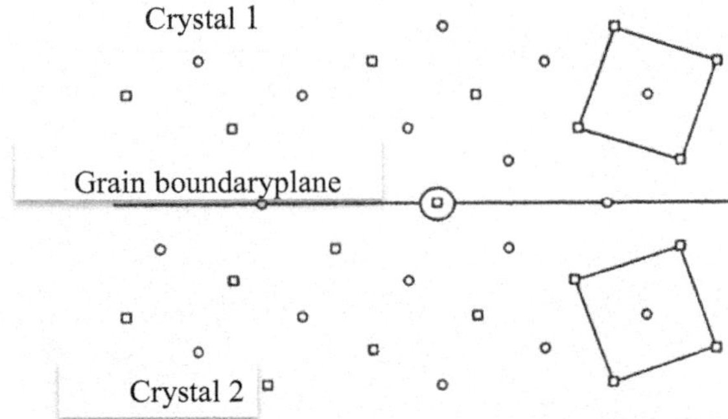

Fig. 2.18 Diagram of a grain boundary clearly showing the different orientations of the lattice in the two crystals; the (slightly larger) atoms in the plane of the boundary are common to both crystals

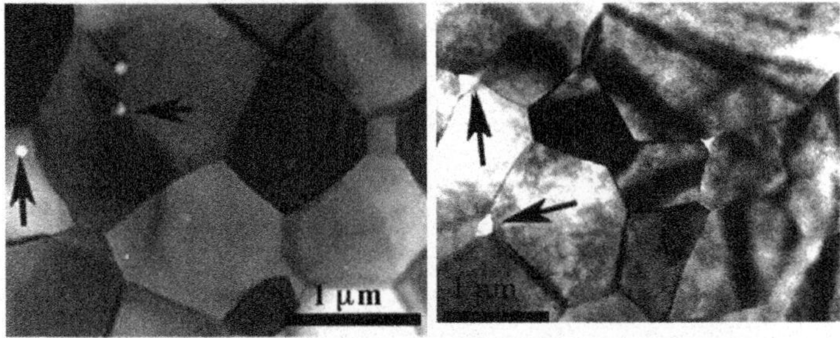

Fig. 2.19 Some pores inside (left) and at grain junctions (right) in alumina (ceramic)

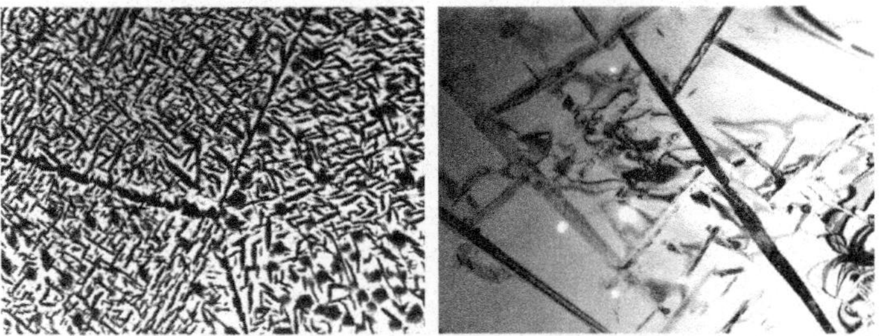

Fig. 2.20 Scanning electron microscopy (left) observation of needle-shaped precipitates within three grains of a titanium/niobium/aluminum ternary alloy. At the grain boundaries, the same precipitates are concentrated in the form of cobblestones. On the right, detail of the needles in transmission electron microscopy

Microstructure of a Crystalline Material

Microstructure reflects the organization of atoms in the solid, including local changes in the nature and position of some of them.

Literally, *structure at the micron scale* (1 micron = 1 millionth of a meter), microstructure is more generally used to describe the fine structure of a solid, including its defects. It covers dimensions ranging from a few nanometers to a few centimeters. The term *structure* refers to the overall arrangement of atoms, whether ordered or disordered. The structure often varies locally, making knowledge of the microstructure necessary.

A crystalline material is composed of several crystals. The microstructure integrates the internal defects of each crystal, their nature, density, and distribution, as well as those characteristic of the entire polycrystal. A formal distinction is made between geometric and chemical imperfections, but in practice, they are often associated.

The existence of vacancies, always present in a crystal, is indirectly deduced from experiments, and their organization can be simulated. Their concentration increases considerably with temperature. In copper, it is equal to 1/10,000 at a temperature close to the melting point, but in some alloys, it can reach a few percent. Interstitial or substitutional atoms are distributed homogeneously or segregated* on other defects (dislocations and grain boundaries). The density of dislocations in crystals is usually high, generally greater than 1,000 m in length in a 1 cm cube of metal. Dislocations form stacks, tangles, or are organized into microcell walls (Fig. 2.21).

The microstructure of a polycrystal also involves describing the organization of the grains. The morphological texture describes their shapes and dimensions, the crystallographic texture and their orientations (Plates 1, 2).

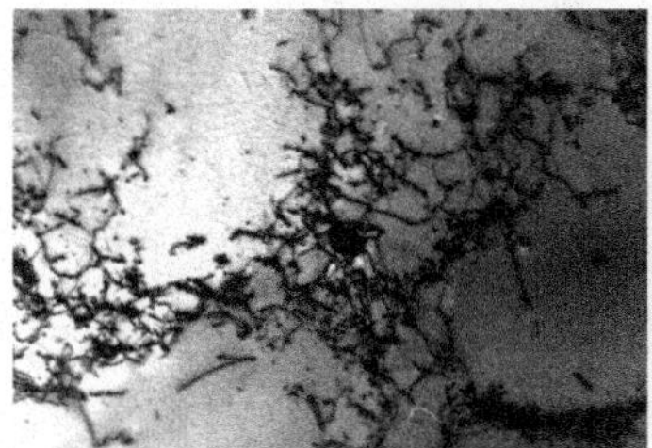
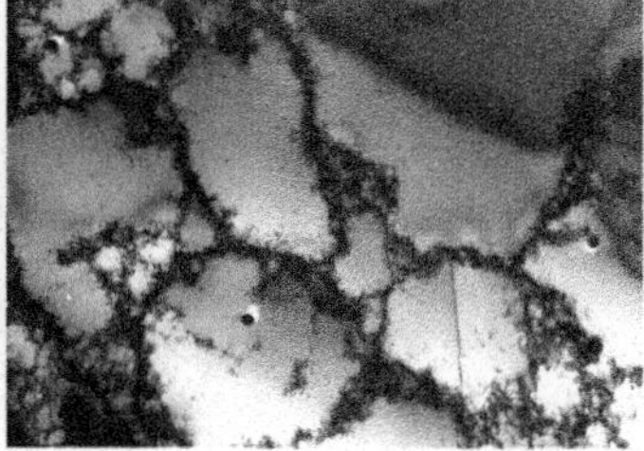

Fig. 2.21 Two dislocation configurations: on the left, entanglement, then, on the right, cell arrangement in copper after 10% deformation

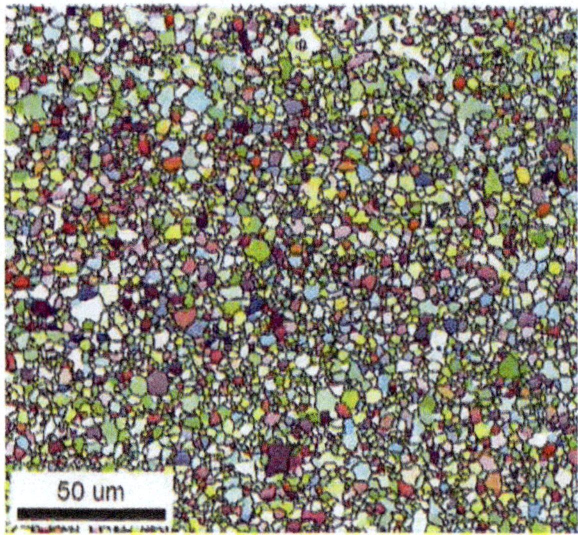

Plate 1 Homogeneous microstructure of a zirconium alloy seen under an optical microscope: the different grains are visible in polarized light. There is no preferred texture

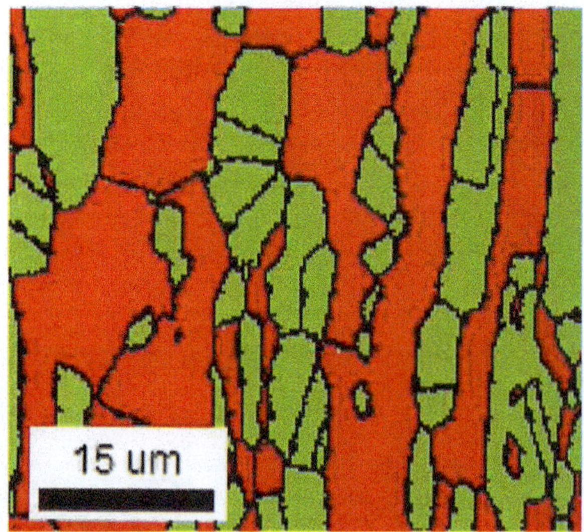

Plate 2 Microstructure observed by scanning electron microscopy showing a non-homogeneous distribution of grains in an iron-nickel alloy. In one band, the grains have almost the same orientation

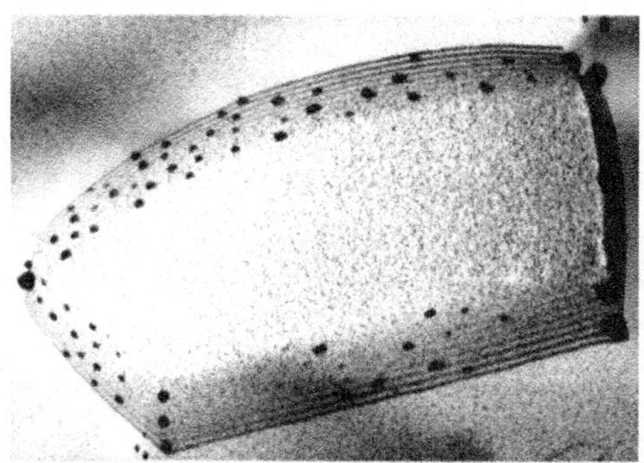

Fig. 2.22 Lead precipitates (in black) at the grain boundaries surrounding an aluminum crystal within a polycrystal; the center of the grain is devoid of precipitates

Finally, the chemical nature of the different crystals, homogeneous or heterogeneous, is of great importance. In particular, we must know the location of the precipitates: intragranular (within the grains) or intergranular (at the grain boundaries) (Fig. 2.22).

The *microstructure* is observed in a *micrograph* (optical or electron photography) after various mechanical and chemical techniques are used to prepare a material *sample*. The "preparation/observation" combination constitutes the method known as *metallography*. The material can thus be described at the mesoscopic, microscopic, and then nanoscopic scales.

While some physical properties are insensitive to microstructure (density, expansion, etc.), the majority are very sensitive. All microstructural parameters can influence properties: the sheet metal used to manufacture a can by stamping* must have a certain crystallographic texture; segregation at grain boundaries leads to very dangerous localized corrosion; small, homogeneously distributed precipitates help harden an alloy; large precipitates of the same type have the opposite effect.

The Unordered State: Amorphous Solids

An amorphous solid is like a frozen liquid. It is a disordered, out-of-equilibrium system that constantly evolves as its ages. Fortunately, this evolution occurs over archaeological periods and does not deprive us of the discovery of prehistoric obsidian knives* and the admiration of the stained-glass windows of Gothic cathedrals.

The amorphous or vitreous state is a metastable* state of matter which, at equilibrium, exists in only three states (crystalline, liquid, or gaseous) depending on the temperature and pressure conditions. Disorder at large distance that characterizes the amorphous structure generally results in homogeneity of the material: one region is statistically identical to another, regardless of the distance between them. However, crystalline domains may exist within the amorphous material, their arrangement being described in terms of microstructure.

Upon cooling, a crystalline material abruptly changes from liquid to solid at a *fixed solidification temperature*. In contrast, amorphous materials continuously change from liquid to solid within a more or less extensive temperature range known as the *glass transition temperature*. Very high viscosity or very rapid cooling leads to the formation of a supercooled (viscous) liquid that only truly becomes solid below the *glass transition temperature* (see Fig. 3.17 in Chap. 3, page 91).

The standard vitreous material is common glass, obtained from molten silica (silicon oxide). Since the melting temperature is high (1,700 °C), a *flux*, most often soda (or lime), is added to lower the melting point. Glass is naturally greenish; the addition of manganese makes it colorless. Adding a small amount of metal or metal oxide during the glassmaking process gives it a color: iron oxides for a green or brown tint, cobalt oxide for blue, and copper oxide for a deep red. These effects are used in the art of stained glass. The introduction of certain oxides into glass also promotes the appearance and growth of crystalline phases, giving rise to glass–ceramics.

The homogeneity of glass is the source of its transparency. No barrier interrupts the path of light. Partially crystallized, it loses this property. However, glass is surpassed in this ability, and with much greater brilliance, by precious stones (diamond, emerald, ruby, sapphire), which are single crystals of minerals.

The name glass was given by extrapolation to other amorphous materials. If a liquid metal alloy is suddenly brought to a low temperature (an *ultra-rapid quenching* process), the atoms do not have time to organize, and an *amorphous alloy* or *metallic glass* is obtained. Some amorphous, non-magnetic steels are used in the construction of stealth submarines (evading all means of detection).

At room temperature and under very high pressure, fullerene (a crystalline form of carbon with 60 atoms per unit cell) (Fig. 2.23) forms an ultra-hard amorphous phase capable of scratching diamonds.

Polymers with very flexible chains also adopt an amorphous structure often described as a "plate of spaghetti" (Fig. 2.24).

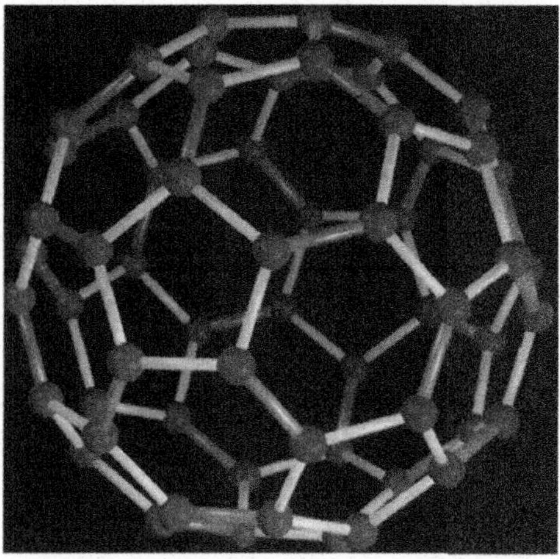

Fig. 2.23 The structure of fullerene consists of 60 carbon atoms arranged at the apex of hexagons or pentagons, in a shape similar to that used to make soccer balls. At room temperature and under very high pressure, fullerene transforms into an ultra-hard amorphous phase capable of scratching diamond

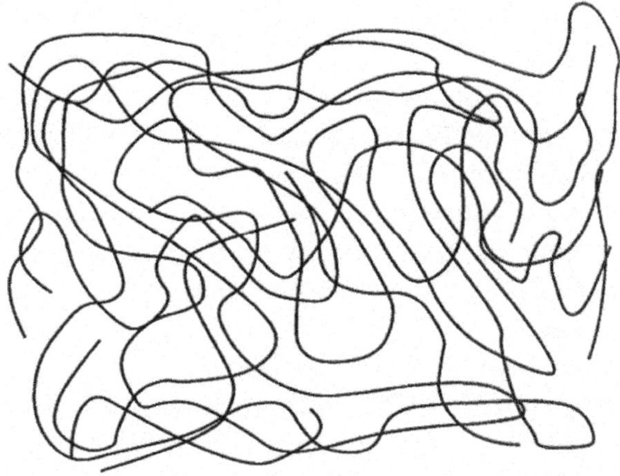

Fig. 2.24 Schematic representation of an amorphous polymer structure

Intermediate-Order States: Polymers

Polymers are most often in a state of intermediate order: far removed from perfect disorder or, conversely, far removed from rigorous order. They form a vital class of materials, the number of which increases every year.

All polymers are obtained by a *polymerization* process, which consists of assembling into large chains an impressive number of elementary molecular units called *monomers*. The dimensions of the chains vary; the *degree of polymerization* is defined as the average number of monomers in the macromolecules (commonly 100 to 10,000). While the bonds within each macromolecule are always covalent, polymers are distinguished by the nature of the bonds between chains.

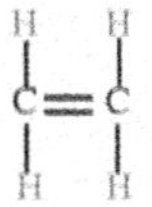

monomer: ethylene

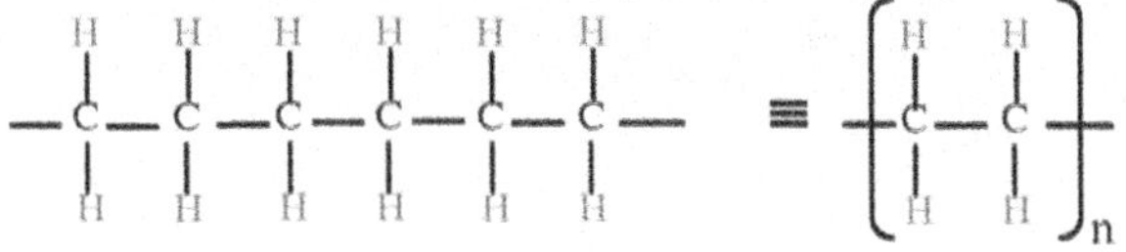

Polymer: polyethylene.

Thermoplastics, partially crystallized or amorphous, have only weak bonds between macromolecules. This is why they soften when heated and are therefore easily molded. Polyethylene (PE) (Fig. 2.25), polystyrene (PS), polyvinyl chloride (PVC), and nylon are widely used or high-performance polymers. Modern humans consume enormous quantities of them: packaging, bottles, inexpensive molded objects, buildings, clothing, etc.

Thermosets contain numerous covalent bonds between chains and are always amorphous in structure. They are produced by mixing a resin and a hardener. The best known are epoxy and polyester. Their use, in the form of Bakelite, Formica, or as a matrix for fiberglass composites, is particularly popular in tableware and household appliances.

With very few covalent bonds, the macromolecules of *elastomers* are flexible and can easily expand and contract back to their original dimensions. These polymers are found in natural and artificial rubbers (tires, seals, etc.). Initially produced from latex extracted from the rubber tree, amorphous rubber is vulcanized (treated with sulfur) and then filled with carbon black to protect it from the harsh sunlight.

Above the glass transition temperature, amorphous polymer chains have great freedom of movement; the materials are then in a *rubbery state*, unlike glasses which are in a supercooled liquid state.

Natural rubber is joined by other natural polymers that are the main constituents of the living world. Crystallized cellulose and amorphous lignin are found in the walls of plant cells; proteins are essential for animal life. In

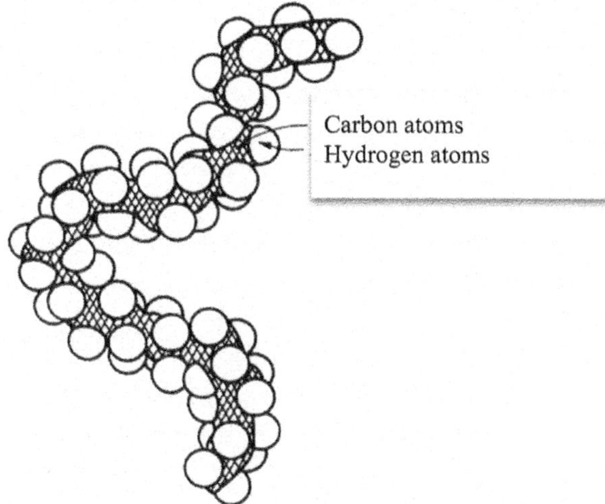

Fig. 2.25 *Three-dimensional appearance of a small part of a polyethylene macromolecule, which is the most popular plastic in the world (packaging bags, cleaning product bottles, toys, etc.)*

particular, collagen, which occurs in the form of a triple helix, is the most abundant protein in the connective tissues of all mammals. It is essential for support, movement, and growth. It is the basis of gelatin. The oldest known glue made from collagen dates back 8,000 years.

The concept of microstructure can be applied to partially crystallized polymers (Fig. 2.26) by taking into account the orientation of the molecule chains and the distribution of small crystalline domains.

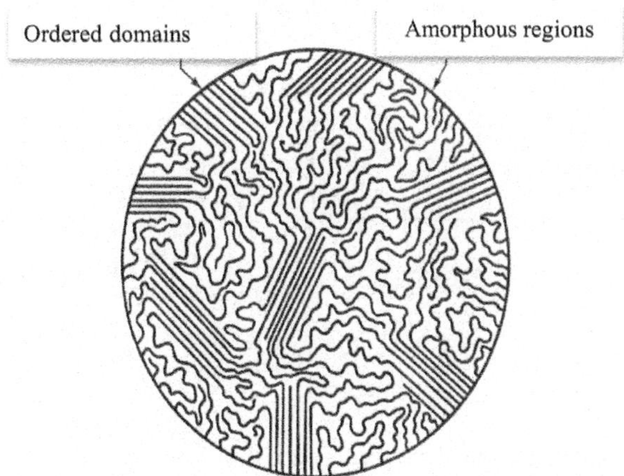

Fig. 2.26 Microstructure of a partially crystallized polymer

Intermediate States of Matter: Liquid Crystals

Liquid crystals are in an intermediate state of matter between solid and liquid states that are nevertheless considered contradictory. Their organization is very sensitive to small temperature variations and to the application of a weak electric field.

A polymer material is in an *intermediate state*, while still being in a traditional solid state of matter. A *liquid crystal* also has a structure that transcends the order/disorder opposition, but above all, it is truly in an *intermediate phase* state*, also known as a *mesomorphic phase*.

The organic molecules that make up a liquid crystal are shaped like elongated rods, more rarely flat disks. They are made of two different parts, both structurally and chemically: a rigid central part and flexible chains at the ends, a hydrophobic* part and a hydrophilic* part. At low temperatures, in the crystallized state, the molecules, parallel to each other, are regularly spaced and form layers perpendicular to their axis. As the temperature increases, the tendency for the molecules to align is counterbalanced by thermal agitation, but, unlike what happens during normal melting, the order only partially disappears; intermediate phases form (Fig. 2.27). Either the layers slide over each other and the molecules are no longer strictly ordered; a certain short-range order, however, remains between molecules within a layer: this is the *smectic mesophase*, which has the consistency of soft soap (from the Greek *smekma* = *soap*). Or there is no longer any assembly in the form of layers, the molecules are distributed in a disorderly manner while remaining aligned in one direction: this is the *nematic mesophase* (from the Greek *néma* = *thread).*

Under the effect of heat, some substances transit through a single mesomorphic state between the crystalline and liquid states; others follow the

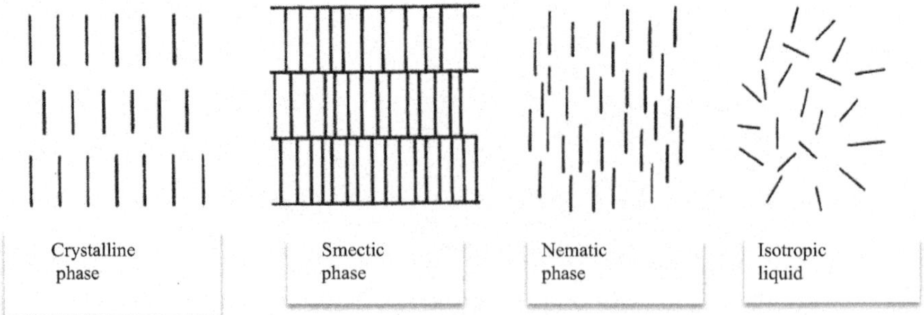

Fig. 2.27 The organization of molecules in the form of rods, in the crystal, the liquid, and the intermediate mesophases found in liquid crystals

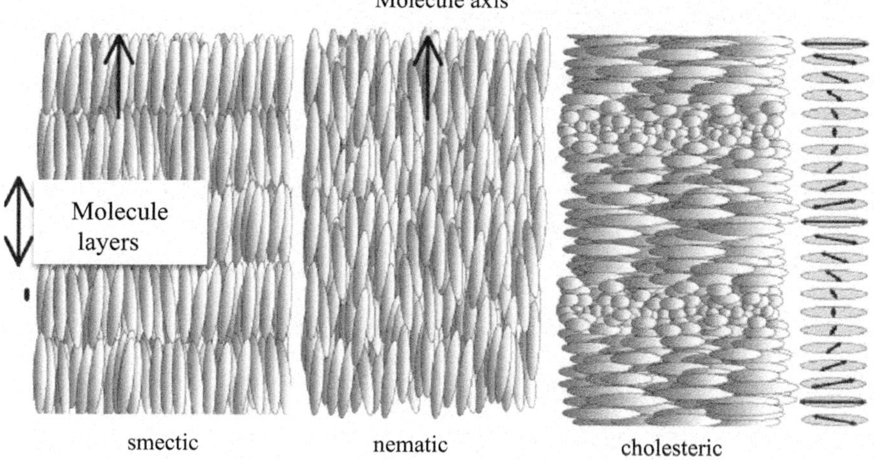

Fig. 2.28 Representation of molecules in a liquid crystal phase: nematic, smectic, cholesteric

sequence crystal → smectic → nematic → isotropic liquid (Fig. 2.28). In all cases, these are true phase changes (see below).

Considered as a variant of the nematic mesophase, the *cholesteric mesophase* has a structure in which the molecular axes are arranged in a helix. Its name comes from the first liquid crystals observed in cholesterol derivatives, a substance known to have harmful effects on our health.

In reality, the orientations of the axes of the various molecules in a nematic crystal constantly fluctuate around the mean direction; a common direction can be imposed on them by an external constraint: an electric field, or, through a surface effect, by placing the liquid crystal in contact with a glass slide scratched in one direction, the molecules then orient themselves along the scratches. These two effects are used in nematic crystal applications.

Mesomorphic phases, unlike liquids, which are isotropic*, exhibit anisotropy that affects their properties. It is the anisotropy of their optical properties that underlies most of their applications. In nematic mesophases, the speed of light propagation depends on its direction relative to the molecules, whether parallel or perpendicular to their axis: this is the phenomenon of *birefringence*. However, applying a very weak electric field to a *liquid crystal* can modify the orientation of its molecules and thus modulate its interaction with light. This effect, discovered around the 1970s, was the origin of the intense development of studies on liquid crystals, although they had been observed since the mid-nineteenth century. It is widely exploited in liquid crystal display technology (see Chap. 4).

Widely present in the living world, cholesteric crystals can appear differently colored depending on whether they are viewed from the front or the side. They also change color with very slight variations in temperature: these characteristics are used in many industrial and everyday sectors (see Chap. 4).

A completely different type of application derives from the mixed hydrophobic/hydrophilic nature of liquid crystals, which is used in biology and cosmetology (soap, gel, etc.).

Alloys, Phases*, and Phase Diagrams*

Water can exist in three states: liquid, gas, and solid. A material also exhibits these different states. Moreover, in the solid state, it exists in different phases* depending on its temperature and composition.

A material is almost never obtained or used in its pure state. The content of additive elements can be extremely low, on the order of one unit per million (ppm), as is the case with dopants in semiconductors. It is generally higher in metals, except for certain applications, and even higher in ceramics, where additives often contribute to the densification of the material. The term *alloy*, originally reserved for metallic mixtures, has now been generalized to ceramics and polymers. A *binary alloy* contains two constituents; for example, basic steels are made of iron and carbon.

The main constituent of the material, or *solvent*, can dissolve a certain amount of another constituent, called a *solute*, to form a *solid solution* similar to the dilution of salt in water, but all in a solid state. In crystalline materials, the solid solution (or alloy) can be substitutional or interstitial, depending on the sites occupied by the solute atoms in the crystal (Figs. 2.14 and 2.15). However, there is a limiting solute concentration beyond which it can no longer be homogeneously distributed in the solvent: this is the *solubility limit*. Another structure appears, in equilibrium with the first (such as solid salt in salt-saturated water), which is called an *intermediate phase**. If this has a precise composition, it is a *defined compound*, but often its composition varies somewhat from the exact composition of the defined compound. An *intermetallic compound* is a defined compound formed of two (or more) metals. In a solid solution of a given composition, atoms of each type can be either randomly arranged (disordered structure) or periodically distributed (ordered structure) (Fig. 2.29).

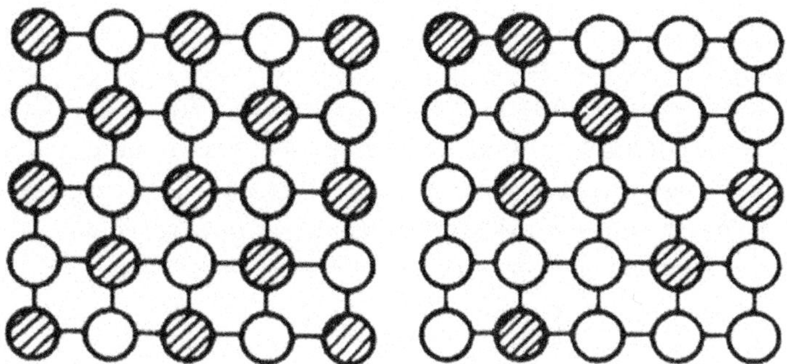

Fig. 2.29 Diagrams of a binary solid solution composed of atoms of different natures: ordered solution (left) and disordered solution (right)

The equilibrium domains of existence of the various phases* of a metal or ceramic alloy as a function of solute concentrations and temperature (pressure being, in this case, kept constant) are plotted on *phase diagrams** (Fig. 2.30).

Ternary (three-dimensional) diagrams account for the phase transformations of a three-component alloy. The dividing lines between phases are replaced by surfaces, making reading the diagram more complex. We often proceed by fixing the composition of one of the components and studying a section of the three-dimensional diagram.

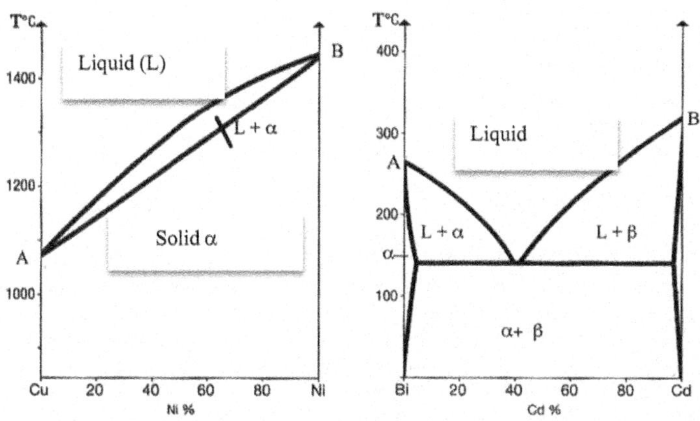

Fig. 2.30 Phase diagrams*. At high temperatures, liquids are completely miscible with each other. The same is true in the solid phase for copper (Cu) and nickel (Ni). Cadmium (Cd) and bismuth (Bi) are poorly soluble in each other; they form very dilute α and β solid solutions, so alloys are generally two phase (α + β). When a liquid alloy is cooled, a two-phase region (liquid + solid solution) exists, except for the pure metal (or oxide) and the defined compound. Points A and B correspond to the melting temperatures of the pure constituents of the alloy

Established at atmospheric pressure, these diagrams are modified if the pressure changes. Alloys that do not exist on Earth are obtained in space laboratories, in zero gravity. Furthermore, *equilibrium phases** are only obtained if cooling is sufficiently slow or the temperature is maintained at a given temperature for a sufficiently long time. Otherwise, during rapid cooling (quenching) or short heat treatment, *non-equilibrium phases** are formed, often with very interesting properties.

Understanding a Material: A Long Journey Through a Vast Space

Controlling and improving the properties of a material requires undertaking a long journey through a space whose dimensions range from those of an atom to those of an ingot*.

The journey into the world of materials takes us through different regions, in ascending order of dimensions: nanostructure, microstructure, mesostructure* (from a few microns to a few millimeters), and macrostructure (visualized with the naked eye). The term *structure* here has the same meaning as the organization of matter as that given to it by physical chemists. The mechanic speaks of *structure* to describe the framework of a vehicle or a building (the concrete structure of a bridge, for example). At each stage of the journey, the region of the material revealed to us is studied using specific methods related to its scale (Fig. 2.31). The organization of the constituent elements (according to the scale: atoms, cell units, crystals, defects, phases, ordered domains, etc.) is described and their chemical composition is analyzed.

(a) ***High-resolution transmission electron microscopy*** *reveals the atomic structure of a grain boundary (indicated by the white lines); the atoms are imaged here as white circles; the simulated image (right) supports the experimental description, and* vice versa.

- *In* ***conventional transmission electron microscopy****, the projection of a boundary plane results in a band with parallel edges on the image; dislocations (shown as white and black lines) are observable in the boundary.*

(b) ***In scanning electron microscopy****, grain boundaries are observed using crystal orientation imaging. The same boundary color indicates the same type of misorientation between neighboring crystals.*

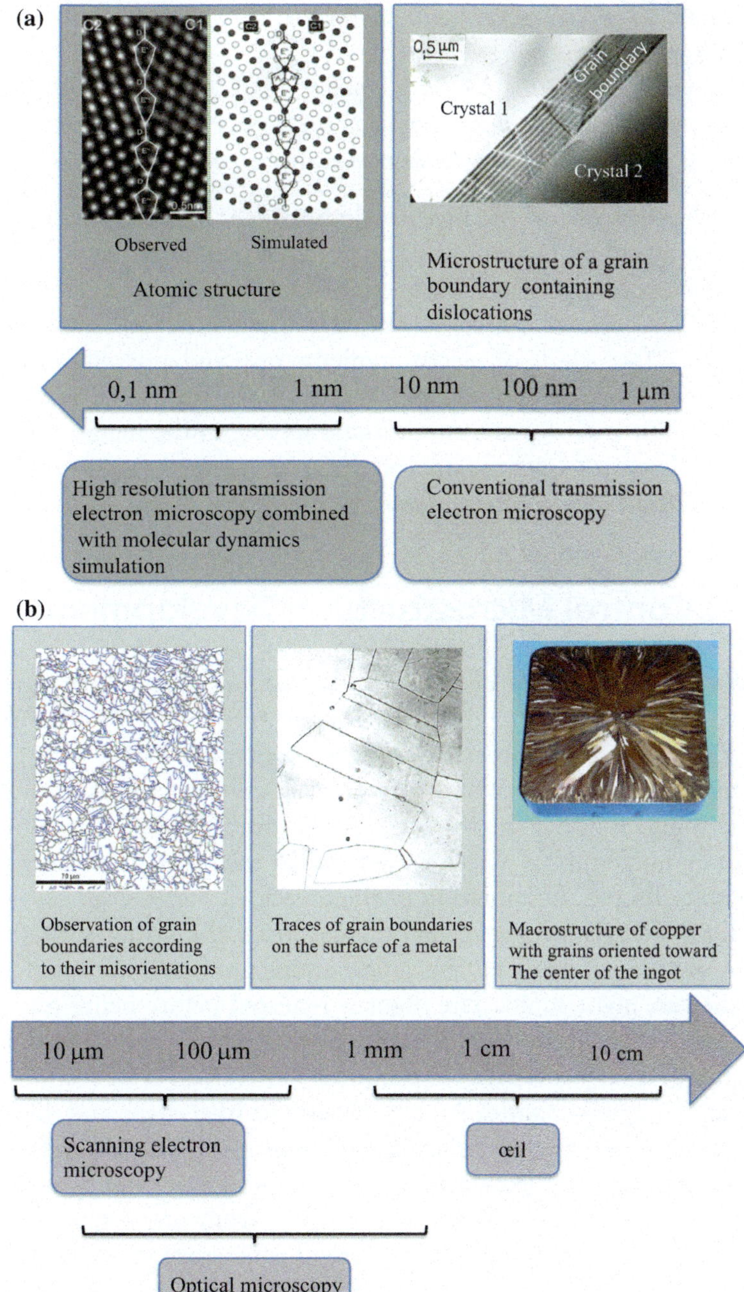

Fig. 2.31 Observation of grain boundaries at different scales (see legend above, page 68, and below, page 70)

- ***In optical microscopy****, grain boundaries are visible by their traces on the free surface of the material.*
- ***By eye****, if the grains are large enough, they can be distinguished by the colors of the oxide, which develops on the surface of the material, depending on the crystal orientation, thus revealing the grain boundaries.*

The *electronic structure,* the first step in our foray into the material, cannot be observed directly; it is deduced using spectroscopic methods and computer simulations. Observations at subsequent scales (nanoscopic, microscopic, and mesoscopic*) require microscopy; they are often grouped under the term *microstructure* (we speak of electron microscopy and not nanoscopy, even when the spatial resolution* reaches 0.1 nm). All the information concerning the microstructure allows us to understand its role in the material's behavior. This approach, which aims to establish the *relationships between microstructure and properties,* is *the foundation of materials science.*

How to Control Microstructure: Development

From the moment it is developed, a material has a microstructure which it will retain, to a greater or lesser extent, even after the modifications caused by various thermomechanical shaping treatments.

Two major stages give a material its final microstructure: its development and its shaping. By final microstructure, we mean that of the material as it is used to manufacture an object. In reality, the microstructure evolves slowly over time, leading to deterioration of the object.

Metallic materials are generally obtained from the liquid state (resulting from the melting of ores) by the *solidification* process. The liquid metal is poured into an ingot mold. The ingot is reheated before being transformed into *semi-finished products,* blanks for the formation of sheets or bars. Continuous casting (Fig. 2.32) is used for products with a small cross section (square, round, etc.) and great lengths. An object, with its final shape and dimensions, can also be manufactured by pouring the liquid metal into a mold (precision or fine art casting). Any solidified metal product has a *solidification texture*: the shape and orientation of the grains differ from one region to another (Fig. 2.33). In addition, the impurity content is heterogeneous, with excess in the regions that solidify last and between crystals (a *segregation* phenomenon).

The crystalline solid most often grows in the form of dendrites, the most well known of which are the star-shaped branches of a snow crystal (Fig. 2.34). The structure of this crystal is reminiscent of that of trees

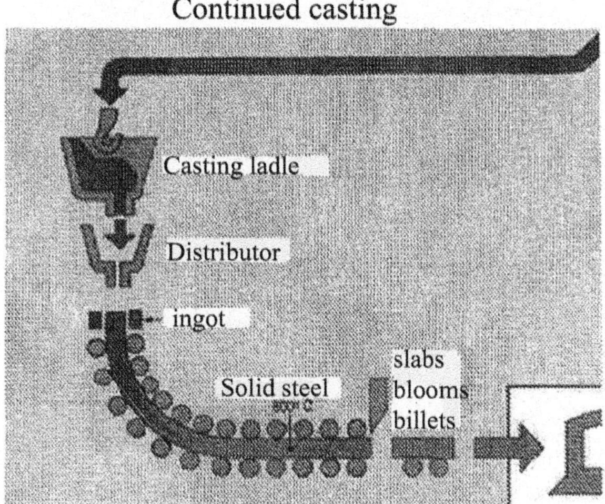

Fig. 2.32 Continuous casting of steel into a copper ingot mold with a square, rectangular, or round cross section. The metal is pulled downward by a set of rollers, solidifying with the help of parietal jets of water

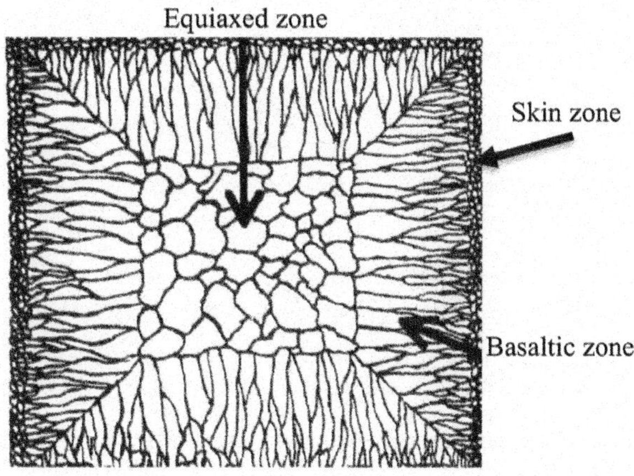

Fig. 2.33 Section of an ingot showing the three characteristic solidification zones: the skin zone with small grains, close to the mold walls, the columnar or basaltic zone with elongated grains normal to the walls, and the equiaxed zone

(dendrite etymologically means "like a tree") (Fig. 2.35). The basalt structure in a metal ingot also results from dendritic growth.

Solidification by very sudden cooling or *rapid quenching*, combined with severe deformation conditions, leads to the formation of nanomaterials (or to amorphous materials).

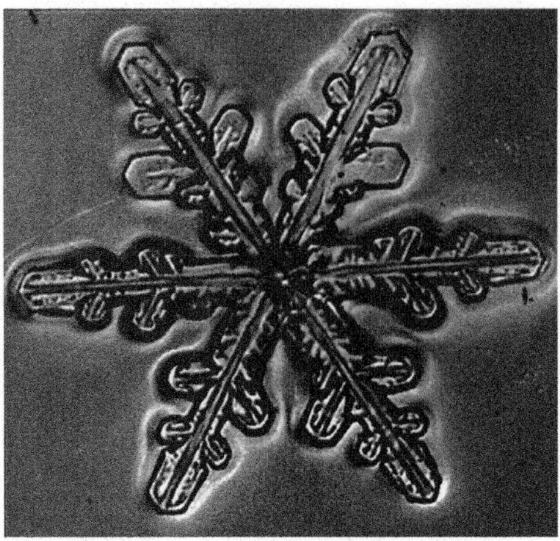

Fig. 2.34 Star-shaped snow crystal, each branch is a primary dendrite with secondary dendrites

Fig. 2.35 Dendrites in an iron-nickel alloy seen under a scanning electron microscope

Glasses can be molded like metals, but the very high melting temperature of ceramics precludes any casting. The sintering process is then used (Fig. 2.36), which consists of compressing powders at room temperature and then annealing them under load, or not. Sintering begins with the formation of bridges between the granules in contact (Fig. 2.37), with the appearance of voids (pores) that partially disappear during heating. The resulting product is more or less dense. Some refractory metals and complex-shaped metal parts are also manufactured by sintering.

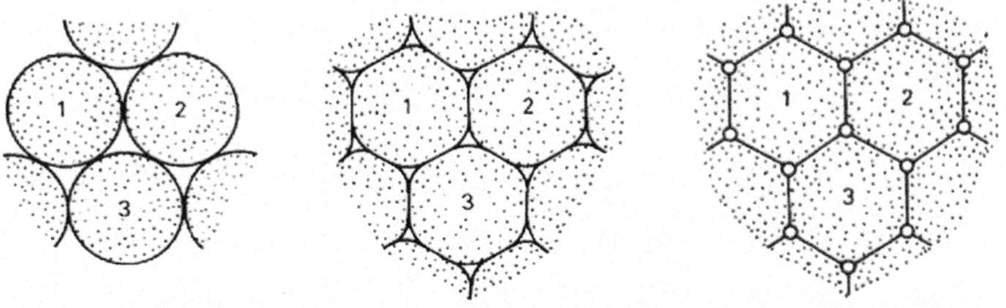

Fig. 2.36 The powder granules sinter under pressure, deforming and reducing the voids (pores) between them. The final structure generally retains small, quasi-spherical porosities

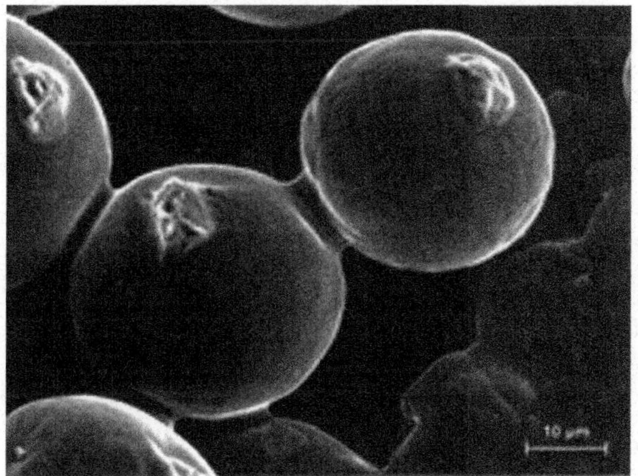

Fig. 2.37 Formation of bridges by sintering stainless steel particles

Adding a very small amount of certain elements called *dopants* to the base ceramic improves density and produces a more homogeneous microstructure (Fig. 2.38).

Plastics are produced by *chemical synthesis* from a base resin. To form a polymer with a large number of macromolecules, an addition reaction initiated by an initiator (in the case of polyethylene) or condensation reaction (in the case of nylon) of monomers is triggered. The *degree of polymerization* depends on the pressure and temperature of the reaction.

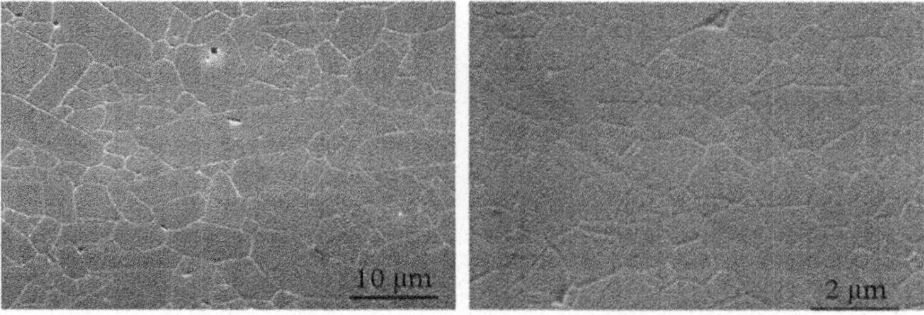

Fig. 2.38 The microstructure after sintering of alumina doped with 0.0005% magnesia (right) is much more homogeneous and much finer-grained than that of pure alumina (left)

How to Manipulate the Microstructure: Shaping

Shaping a material involves performing alternating heating, cooling, and deformation operations to directly give it the shape of an object or an intermediate shape suitable for creating that object.

The shaping of metallic materials after casting is done by cold or hot deformation. *Forging* is a technique, already used by our ancestors, which consists of hammering the metal to give an object a complex shape. In *rolling*, the thickness of a metal block is reduced until a sheet is obtained by crushing it in successive passes between two rolls whose spacing is gradually reduced. *Extrusion* consists of pushing the heated (but not liquid) metal into an orifice (die) with a possibly complex profile to create tubes or bars. *Wire drawing* transforms the metal into wire by passing it through a die under the effect of traction. Finally, a flat sheet of metal is deformed using a *stamping* punch to obtain a part of a specific shape: this is the process for shaping cans.

In all cases, the metal is *plastically deformed.* It undergoes *work hardening*, which results in the introduction of numerous dislocations into its microstructure. If the strain rate becomes too high, the metal risks breaking. Continuing the forming process then requires annealing treatments. *Restoration* allows the dislocations, homogeneously distributed throughout the work-hardened metal, to organize into cells, reducing stresses. *Recrystallization* occurs through the nucleation of new, undeformed grains that gradually replace the deformed grains in the microstructure. The *annealed metal* can then be *work hardened* again. Thus, *work hardened* and *annealed* refer to two opposing but complementary states: an annealed metal is work hardened, and a work-hardened metal is annealed (Fig. 2.39).

Recrystallization annealing also serves to control grain size, given that manufacturers almost always seek fine-grained metal. Another annealing

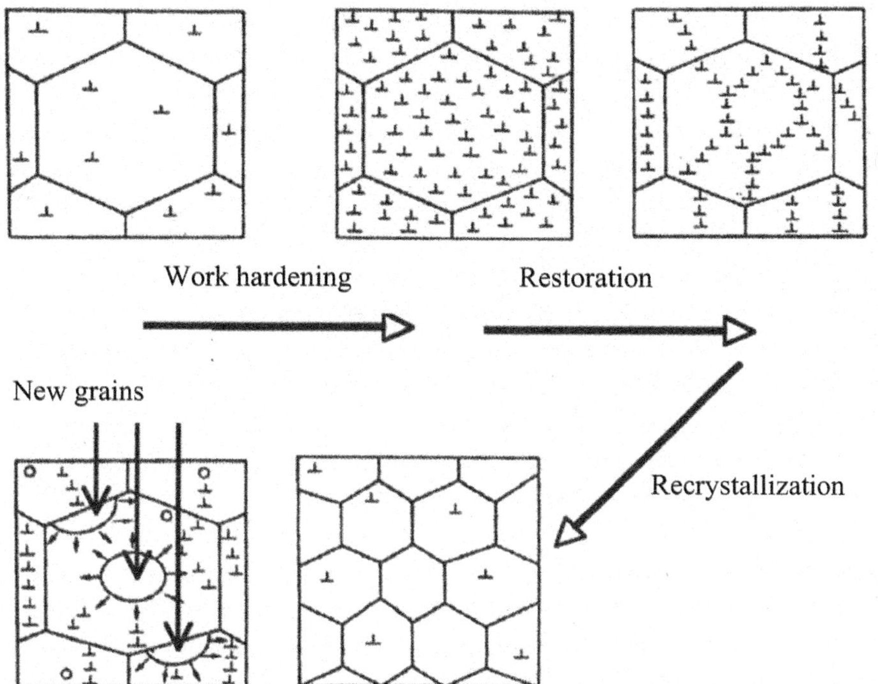

Fig. 2.39 Diagrams showing how the microstructure of a metal evolves from the work-hardened state to the annealed state, with reorganization of dislocations (⊥) and the appearance of new crystal nuclei

process, called *homogenization annealing*, eliminates the compositional heterogeneities of an ingot resulting from its solidification. *Thermomechanical treatments*, alternating deformation and annealing, result in surprisingly different microstructures for the same material (Fig. 2.40), associated with equally very different properties.

Glasses also undergo hot pressing (equivalent to forging) or rolling (windows). Two shaping techniques are characteristic of them: blowing (bottles, light bulbs, etc.) and flotation, a process invented in the mid-twentieth century that involves continuously floating a ribbon of glass on molten tin (glass for windows and mirrors). Terracotta is molded, turned, or shaped, then dried before being fired in kilns.

The most common and least expensive method for shaping polymers is *extrusion*. Similar to a pastry nozzle, it involves pushing the fluidized polymer through a die and then cooling it to maintain the desired shape: tube, ribbon,

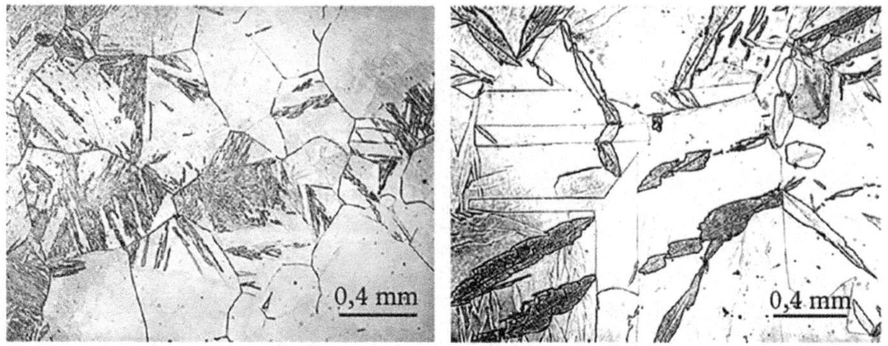

Fig. 2.40 The same 30% nickel iron alloy annealed at the same high temperature exhibits two different microstructures depending on whether it is cooled slowly (left) or abruptly to—60 °C (quenching) on the right. Under optical microscopy, the secondary phase (in gray) exhibits a different morphology and distribution within the initial phase (white)

etc. This is a process similar to metal wire drawing. Thermoplastic polymers, which soften when heated, are shaped by molding. For thermosets and elastomers, heating, shaping, and curing occur simultaneously during compression molding.

How to Play with Shapes: Tailor-Made Materials

Designing a tailor-made material not only involves combining different materials while controlling the microstructure of each, but also optimizing the geometry of each component. This is how we move from composites to hybrid materials.

Composites, intimate blends of two or more homogeneous materials at the finished product level, constitute the first attempts at "tailor-made materials." The more recent development of *hybrid materials* better meets often contradictory specifications. It is an assembly of solid materials juxtaposed and welded together in a single part.

A composite is composed of a *reinforcing material* dispersed in the form of fibers or particles in a *matrix* that surrounds them. The reinforcement provides mechanical strength, while the matrix ensures the cohesion of the structure and the transmission of forces. Furthermore, between the reinforcement and the matrix, there is an interface whose characteristics are fundamental to the material's strength. *Fiber composites* (Fig. 2.41) are composed largely of glass or carbon fibers, long or short, embedded in a

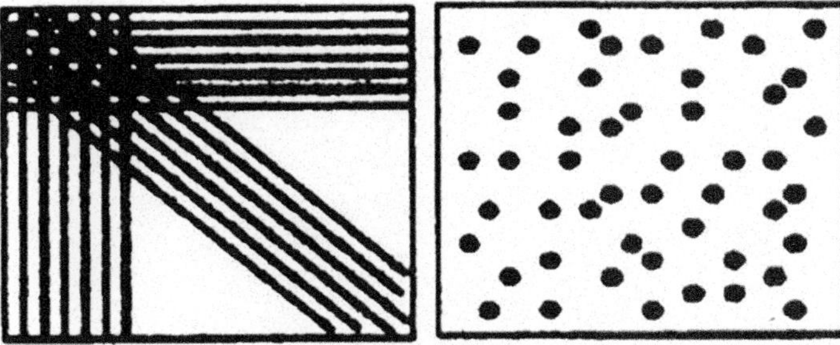

Fig. 2.41 Schematic representation of a fibrous composite material (left), particulate (right). The particles are evenly distributed throughout the material. This results in better tensile strength and a better strength-to-weight ratio

polymer matrix. Small, randomly arranged short fibers do not have as significant an effect as long fibers arranged parallel to each other, but they can be organized during molding, as is the case for tennis racket frames. Fiber composites are heterogeneous and anisotropic* materials; precise knowledge of the mechanical stresses to which a part will be subjected is required before deciding on their use. *Particulate composites* (Fig. 2.41) are generally formed from a metal and fine ceramic particles. Thereby, the dispersion of tungsten carbides in cobalt produces a composite called *cermet*, which is very strong and used for cutting tools.

A major advance in the design of "custom-made materials" has emerged with *hybrid materials*: not only do we play with the microstructures of the components, but also with their shape. Sandwich materials are being created, among other things: a structure formed by a layer of high-performance cardboard (or epoxy resin) bonded between two layers of carbon fiber-reinforced polymer fairly closely reproduces the performance of a wooden soundboard. It recreates the timbre qualities of a violin, even if you're not playing a Stradivarius!

Material design is often guided by knowledge of the organization of natural materials, whose richness comes from the variety of forms at all scales and from the multiplicity of interfaces. In particular, biological systems, often composed of organic and inorganic materials, are capable of organizing the mineral phase into highly elaborate hierarchical structures (Fig. 2.42).

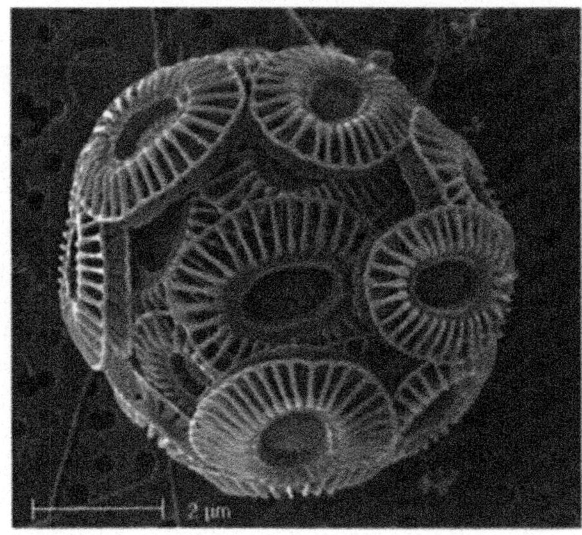

Fig. 2.42 A coccosphere is a unicellular algae surrounded by an assembly of calcareous platelets called coccoliths, the whole structure having a diameter of 10 µm. When the algae die, the coccoliths dissociate and settle on the seabed, forming thick layers of sediment. The natural chalk cliffs in Normandy are essentially composed of hundreds of billions of coccoliths

3

The "Life" of Materials

Metals have been personalized since the Sumerian era and this comparison persists. We talk about aging, fatigue, and the lifespan of materials. Indeed, and contrary to what is perceived from the outside, a material is not inert. Its atoms displace, attract or repel each other, unite or annihilate. It is this internal "life" that gives the material most of its functional properties.

Thermal Agitation and Atomic Diffusion

In most solids, atoms vibrate; they move under the effect of temperature and/or mechanical stress. Diffusion, i.e., migration of atoms, controls the speed at which a large number of processes occur.

Each atom in a crystal vibrates on either side of its average position in the crystal lattice. At room temperature, the amplitude of the vibration is, on average, equal to five percent of the distance between atoms (on the order of a billionth of a centimeter); it increases with temperature. But the atoms in a crystal are not independent of each other. A small displacement of one atom is transmitted to its neighbors, and the vibrations thus propagate throughout the crystal. *Thermal agitation* is theoretically zero only at absolute zero. These vibrations are not symmetrical; their amplitude is greater in the direction of separation of the atoms than in the direction of their convergence. This results in atoms moving away from each other as the temperature increases, causing *thermal expansion* of solids.

When thermal agitation increases, *point defects* (see Chap. 2, page 52), always present in crystals, migrate: this is the phenomenon of *diffusion.*

L. Priester, *Materials: History, Science and Perspectives*,
https://doi.org/10.1007/978-3-032-15754-6_3

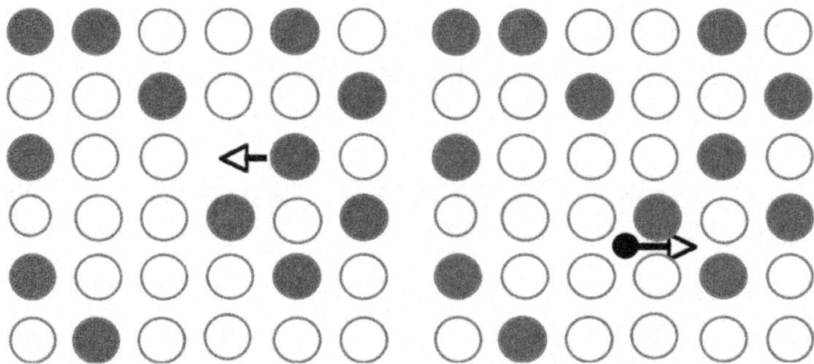

Fig. 3.1 The simplest diffusion mechanism is the jumping of an atom onto a vacancy site (a hole). However, an interstitial atom (in black) can migrate from its site to the neighboring interstice in the crystal

An atom only easily changes position if there is a vacancy around it in which it can move without disturbing its environment too much. A vacancy then appears at the site the atom has just left; the total number of vacancies is conserved. Diffusion is essentially a *vacancy mechanism* (Fig. 3.1). A simple exchange between two neighboring atoms requires too much energy. However, an interstitial atom can jump from one interstice to another neighboring interstice in the crystal.

The migration of an atom in a lattice composed of the same atoms is called *self-diffusion*. If the diffusing species differs from that of the lattice, it is called *hetero-diffusion*. To study it, diffusion couples composed of two metals bonded together are heated long enough for the elements to interpenetrate. The distribution of atoms on both sides of the initial interface is then analyzed (Fig. 3.2). The introduction of atoms of one of the elements in radioactive form or *radiotracer*, whose positions are detected by the radiation they emit, allows the trajectories of the atoms to be tracked. The location of radiotracers in the material after diffusion annealing is determined by autoradiography.

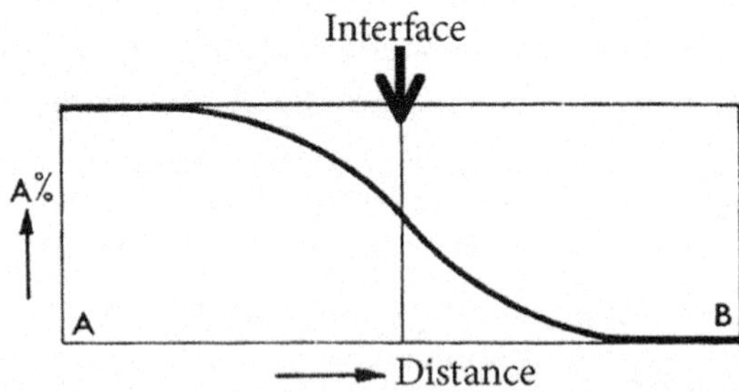

Fig. 3.2 Schematic diagram of the percentage of element A in the A/B diffusion pair after annealing

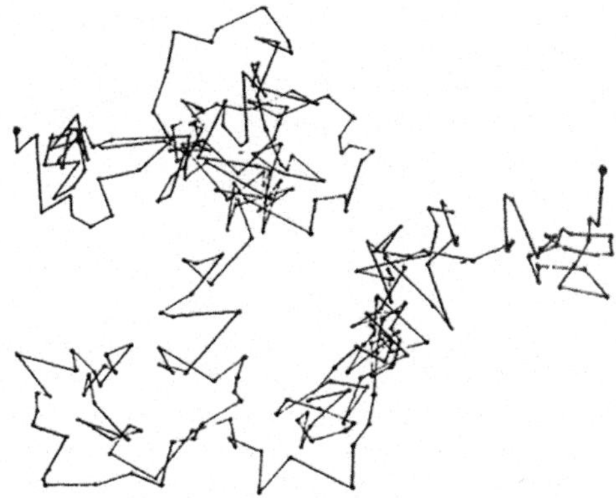

Fig. 3.3 Random walk (or Brownian motion): trajectory of 0.4 μm diameter gum resin particles in aqueous suspension, located at equal time intervals

The trajectory of an atom in a crystal is random. The best known example of a random walk is the Brownian motion of small particles suspended in a liquid, discovered by Robert Brown in 1828 (Fig. 3.3).

In iron at 25 °C, a carbon atom spends an average of one second at a site before making a jump. This is also the residence time of a copper atom at a site at 400 °C, but if the temperature reaches 850 °C, this time drops to one microsecond (one millionth of a second).

Diffusion along dislocation lines and, even more so, at grain boundaries is significantly faster (a thousand to ten million times faster depending on the temperature) than in the bulk of the crystal (Fig. 3.4); these regions are therefore called *diffusion short-circuits* or *pipe-diffusion* for dislocations.

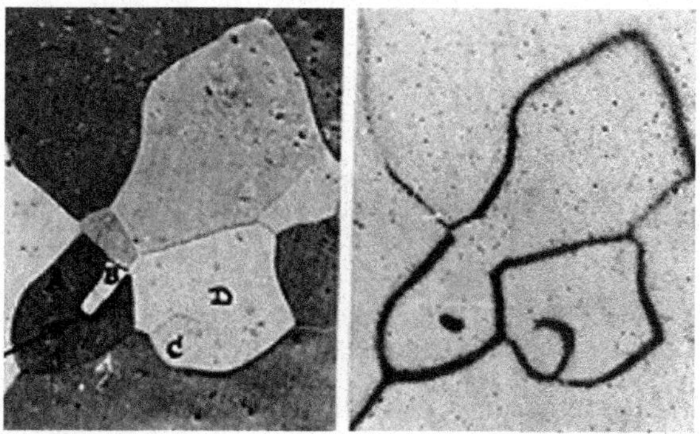

Fig. 3.4 The different grains of an iron sample are visible by their grayscale (left). After deposition of radioactive iron on the surface and diffusion annealing*, autoradiography (right) taken at a certain depth from the surface shows selective penetration of radioactive iron into almost all the boundaries (which appear as thick black lines); few boundaries have resisted to penetration

In crystalline materials, the phenomenon of diffusion is well understood; it governs the kinetics of most processes: phase transformations, plasticity of crystalline solids, and high-temperature oxidation. Diffusion also occurs in amorphous materials and polymers, but its mechanisms are poorly understood.

Oxidation and Corrosion—A Single Fundamental Reaction

The atoms of the solid react with those of the surrounding medium (liquid or gas), most often leading to degradation of the material. However, some reactions protect the surface of the material or improve its appearance.

Metals in contact with a gas or liquid tend to form compounds (oxides, carbides, sulfides), stable forms in which they are present in most ores. Corrosion is called oxidation in an aqueous medium, while the term oxidation is reserved for the reaction in a gaseous medium, also called dry corrosion. In both cases, the reaction between the atoms of the solid and those of the aggressive medium requires diffusion to bring the species into contact.

Corrosion of a metal leads to the formation of *positive ions** or *cations*, dissolved in the liquid, and free electrons that react with water and oxygen dissolved in the water to form *negative ions** or *anions*. Cations and anions form oxides or hydroxides*. Corrosion, which occurs *uniformly* over the entire surface of the solid, is often not very dangerous. On certain metals, it quickly leads to the formation of a protective oxide or passive layer (alumina film on aluminum) which significantly reduces the corrosion rate without totally eliminating it. Atmospheric corrosion, aqueous because air always contains moisture, is uniform corrosion. However, it proves very harmful when the air is polluted by products from oil combustion or by chlorides (marine atmosphere).

Much more serious is *localized corrosion* (Fig. 3.5), which results from local heterogeneities of the surface. This surface presents specific defects related to its preparation (embedded particles, scratches, etc.) or holes in the passive layer. But above all, the surface is a site where defects in the material's microstructure (dislocations, grain boundaries) emerge, often associated with foreign atoms, making them prime sites for attack. Localized corrosion occurs through *pitting* (Fig. 3.5) or *crevices* (marine corrosion). Finally, *intergranular corrosion* can lead to complete decohesion of the grains (Fig. 3.6). Corrosion is generally aggravated in the presence of stresses within the material.

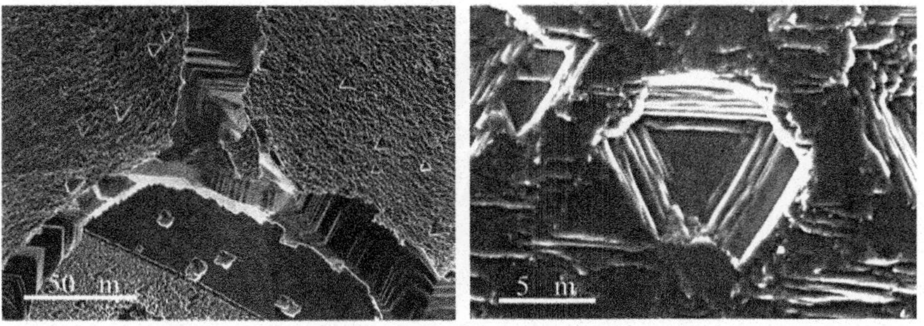

Fig. 3.5 Grain boundary corrosion (left) and pitting corrosion (right) of pure nickel in sulfuric acid, showing that it propagates in steps

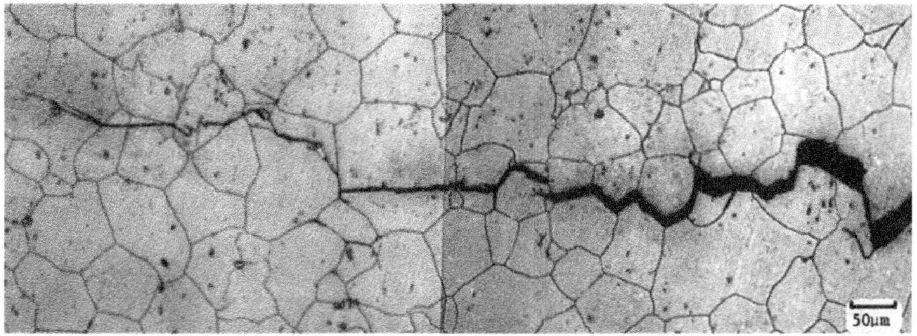

Fig. 3.6 Stress corrosion in sulfuric acid of a nickel–iron-chromium alloy. The attack propagates (from right to left) mainly following the grain boundaries

Oxidation is the reaction of a solid with a gas, typically atmospheric oxygen, but also chlorine, sulfur, nitrogen, etc. In the case of oxygen, a surface oxide film forms. The reaction continues if the surrounding oxygen reaches the metal or if the metal comes into contact with the atmosphere by diffusion through the thin oxide layer already formed. In the very early stages of oxidation, the oxide crystallites may appear oriented on the metal's surface (Fig. 3.7).

In some cases, oxides of different types form as layers superimposed on the surface of the metal. In alloys, each element reacts differently with oxygen, giving rise to a mixture of oxides, miscible or immiscible, on the surface of the material (Fig. 3.8).

All materials undergo degradation related to their environment: concrete in seawater, plastics in the presence of oxygen (risk of combustion at high temperatures) or in contact with organic solvents, and ceramics in contact with dairy products* (mainly silicate).

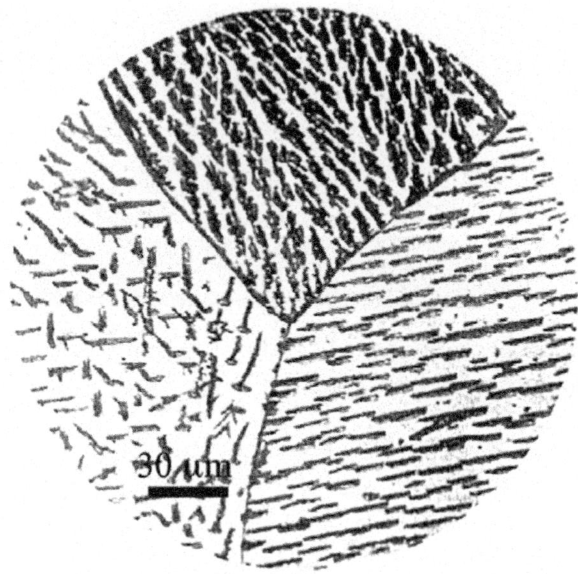

Fig. 3.7 The first copper oxide particles that develop on three copper grains have different shapes from one grain to the next, thus revealing the different crystalline orientations of the surface

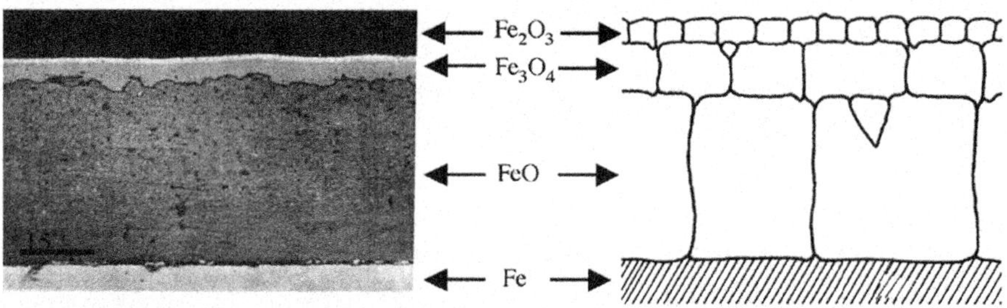

Fig. 3.8 A cross section observed under optical microscopy shows the oxide film formed on pure iron oxidized in air at 700 °C; a diagram showing the microstructure of the different oxides superimposed on the metal

Segregation: A Universal Phenomenon

Atoms do not choose their positions in the microstructure, they are imposed on them by elastic or chemical forces. They are often attracted to crystal defects.

Mainly for reasons of size, but also because of mutual affinity, solute atoms (elements added deliberately) or impurities tend to congregate at certain sites in the microstructure: this is the phenomenon of *segregation.* Defects are preferential sites for segregation: indeed, the space available for an atom near a dislocation line, and even more so in a grain boundary, is generally more comfortable than within the crystal.

It is difficult to visualize the segregation phenomenon because, unlike precipitation*, there are no new phases observable by electron microscopy, but simply a local enrichment (or conversely, a depletion) of atoms of one of the species composing the material. These compositional differences can be detected by chemical analyses at the micrometer and nanometer scales. They also cause different corrosion sensitivities between different regions of a material, which can be used to indirectly reveal the presence of segregation through chemical attack (Fig. 3.9).

Computer simulation of the atomic structure of a crystal defect makes it possible to predict the segregation sites of a given species. In a grain boundary, a monoatomic layer of segregated elements often forms between the two crystals (Fig. 3.10). Only one technique can visualize the local overconcentration (or depletion) of atoms of a species: atom probe tomography* (Plate 1).

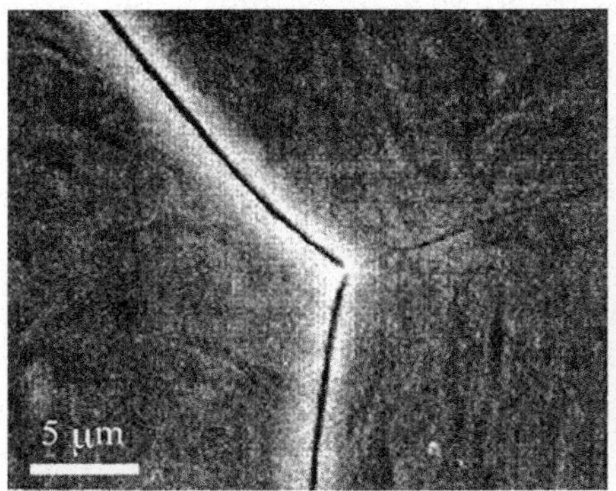

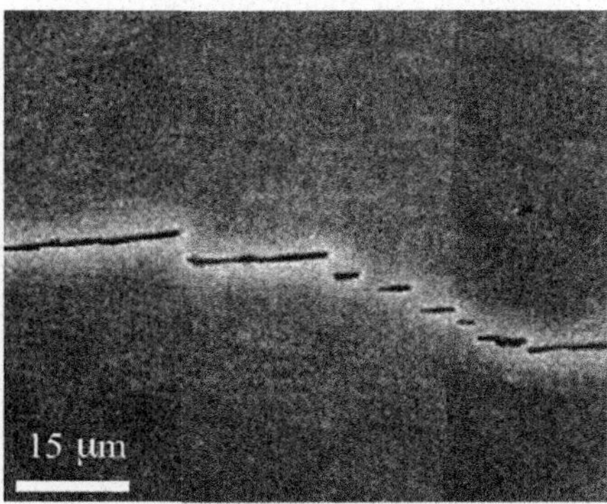

Fig. 3.9 Electron images showing the more or less strong segregation of phosphorus in three grain boundaries (left) or in certain facets of the same boundary (right) in an iron-based alloy with 0.06% phosphorus. Segregation is revealed (black lines) by the preferential chemical attack of phosphorus-rich regions

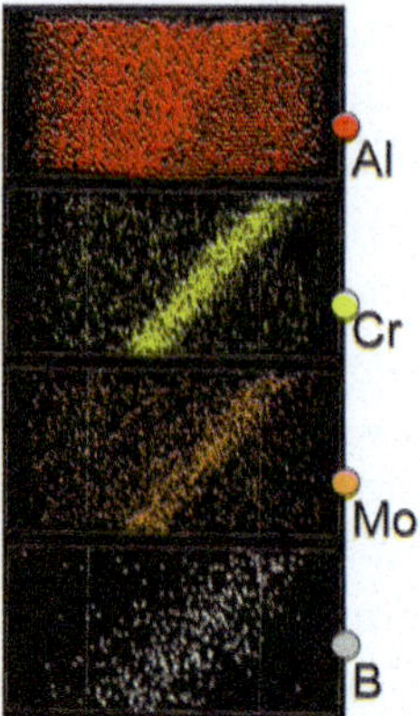

Plate 1 Three-dimensional image, obtained using atom probe tomography, of the distribution of different atoms in a nickel-based superalloy. Each point corresponds to an atom of chromium (in yellow), molybdenum (in orange) and boron (in white). These atoms are segregated over a thickness of approximately 1 nanometer in a grain boundary (whose plane is normal to the figure). Aluminum (in red) does not appear to be in excess in the boundary.

In alloys containing more than two constituents, an element that migrates to a preferred site often pulls along another with which it has a high chemical affinity. For example, carbon segregates at the grain boundaries of certain steels, surrounding itself with atoms of chromium, titanium, or niobium, depending on the alloy's composition. This is called *co-segregation*.

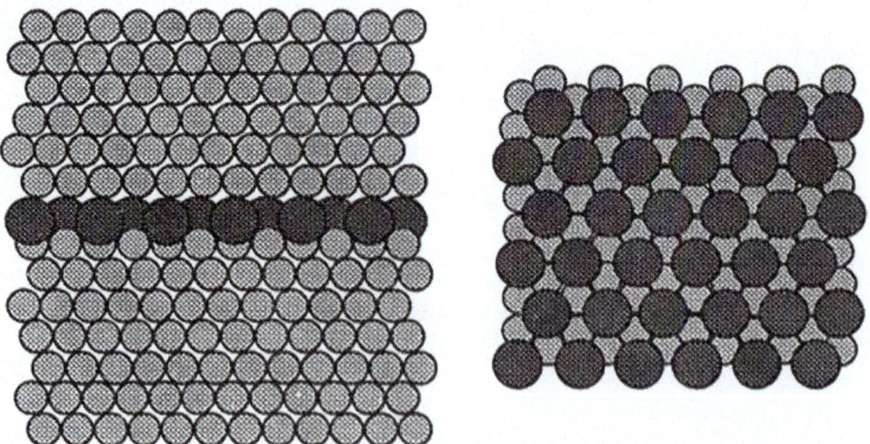

Fig. 3.10 A layer of bismuth atoms (dark gray) has formed by segregation at the grain boundary between two copper crystals (light gray atoms). Results of computer calculations are presented in section and in the boundary plane

The overconcentration of one or more elements in a region of a material can lead to the appearance of a new phase rich in that element(s). Segregation is a phenomenon that often precedes precipitation. In the case of steels, where carbon is located, carbides of chromium, titanium, and niobium appear.

Intergranular segregation often results in reduced cohesion of grain boundaries, which manifests itself as greater sensitivity to fracture and/or corrosion. More rarely, certain elements strengthen the grain boundaries of certain alloys or protect them against chemical attack.

The free surface of a part is also a site of segregation of atoms in solid solution, which modifies its reaction with the surrounding environment. Thus, the carbon present in a large number of alloys rises to the surface, resulting in surface properties that differ greatly from those of the core of the material.

Phase Transformations in a Crystalline Material

Atoms in solids associate differently depending on the temperature and are organized into various lattices. These changes are slow if they occur by diffusion, but can be instantaneous if they result from lattice shear.

Phase transformations involve movements of atoms, arranged differently relative to each other depending on the phase in which they are engaged. Two types of phase transformations exist, described with reference to the movements of humans:

- Either atoms move by diffusion, independently of each other, along varied paths, changing neighbors. These are called *civil transformations.*
- Either the atoms move collectively, as a block, guarding their close neighbors, like an army! These are called *military transformations.*

Civil or diffusive transformations proceed in two stages: *nucleation*, then *growth* of the new phase.

Nuclei are clusters of a few dozen to a few hundred atoms of the new phase; they are only visible under high-resolution electron microscopy*, a technique with a resolution of up to 0.05 nm. They rarely appear *homogeneously*, i.e., dispersed throughout the original phase, also known as the *parent phase*. More often, they are located on crystal defects, dislocations, and grain boundaries, or around precipitates already existing in the alloy: *heterogeneous nucleation* prevails (Fig. 3.11).

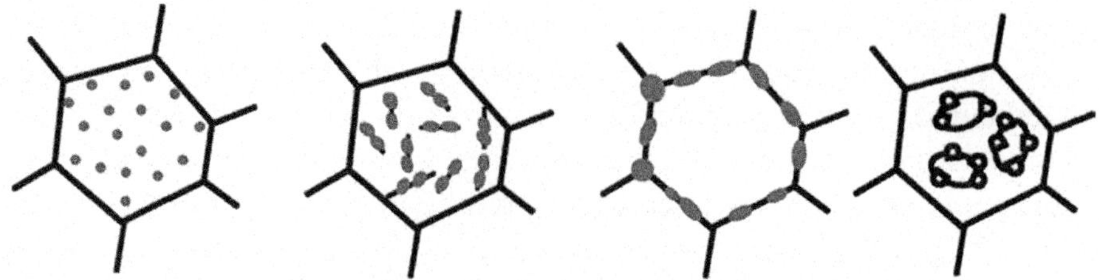

Fig. 3.11 Diagrams showing second-phase nuclei (in gray), either dispersed homogeneously throughout a crystal or heterogeneously arranged on dislocations (small black lines), grain boundaries, or around precipitates

If a good match exists at the interface between the distances of atoms in the new phase and in the initial phase, the interface is said to be *coherent* (Fig. 3.12). This match may not be perfect, in which case the crystals tolerate a small elastic deformation to preserve the connection. If the difference between the interatomic distances on either side of the interface becomes

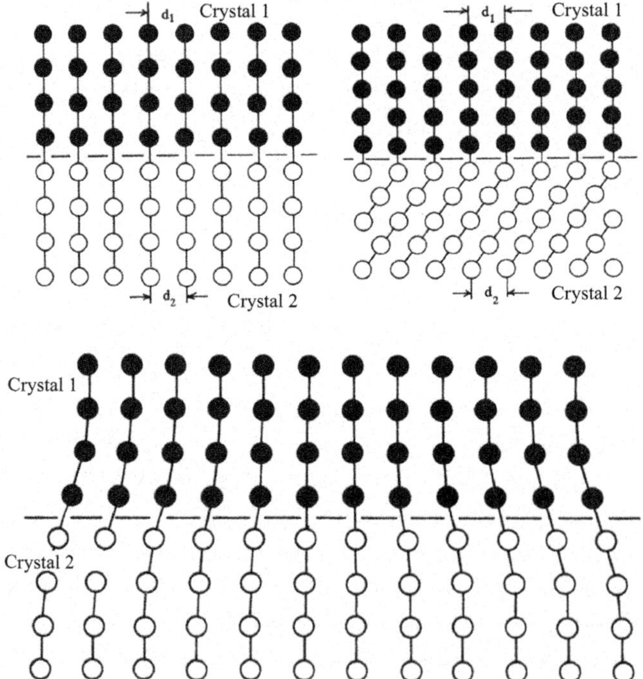

Fig. 3.12 Coherent interfaces: the distances between the atoms of crystal 1 and those of crystal 2 are equal (diagrams above) or nearly equal (diagram below) at the interface. In the latter case, plastic deformation of the crystals occurs on both sides of the interface

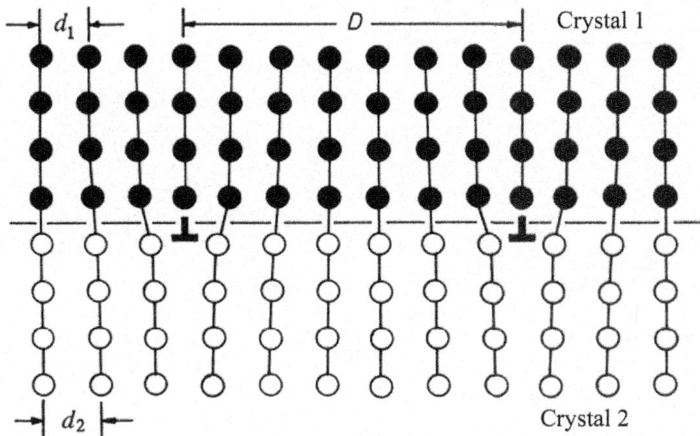

Fig. 3.13 Semi-coherent interface: the distances between atoms in the two crystals differ too much, and additional planes appear in one crystal. The dislocations (⊥) that form at the interface are periodically distributed

too great, additional planes necessarily appear in a crystal; *accommodation dislocations* are distributed periodically in the interface, which is then called *semi-coherent* (Fig. 3.13), because between the dislocations, the rows of the two crystals connect. If no match exists, the interface is arbitrary or *incoherent.*

Depending on the type of interface that border it, a nucleus takes on different shapes. If all interfaces are incoherent, it is roughly spherical within the crystal and takes the shape of an ellipsoid* at a grain boundary. If one or more coherent or semi-coherent interfaces form, the nucleus has one or more planar facets (Fig. 3.14).

A nucleus develops over time to produce a precipitate that retains its initial shape in the early stages of growth (Fig. 3.15), but often loses its faceted appearance as it grows. Growth requires additional atoms to attach to the interface for it to progress to one side or the other.

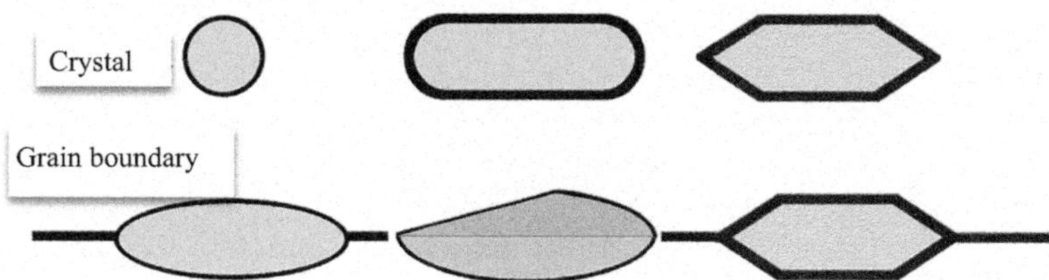

Fig. 3.14 Schematic representation of different seed shapes in a crystal and at a grain boundary, from left to right: all-incoherent, coherent and incoherent, all-coherent and/or semi-coherent interfaces

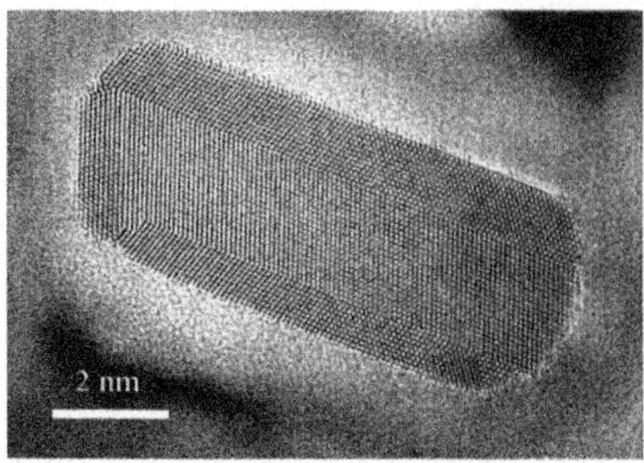

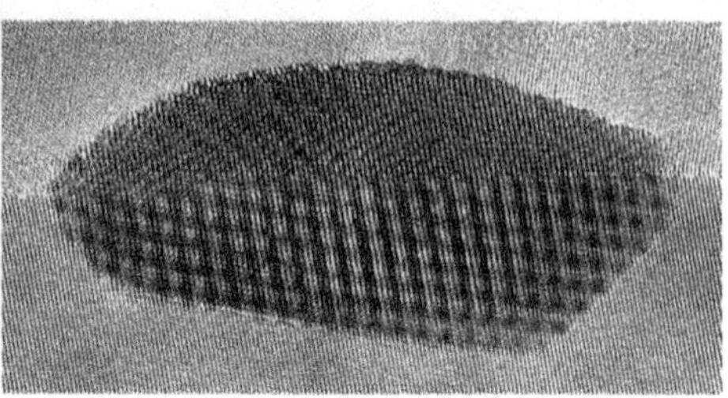

Fig. 3.15 High-resolution transmission electron micrographs showing very small precipitates in the shape of their initial seeds: a germanium crystal within an aluminum crystal (left) and a lead crystal at the grain boundary of a thin aluminum film (right). The flat and curved interfaces are coherent and incoherent, respectively

Military or displacive transformations generally occur through a *quenching* treatment, with the cooling rate being so rapid that it precludes any possibility of atom diffusion. The phase that forms is not the one predicted by equilibrium; it is metastable. Unlike a phase formed by diffusion, it appears, without a change in composition, through shearing of the parent phase lattice induced either by thermal quenching stresses or by mechanical stresses. The nuclei, in the form of ellipsoids*, appear preferentially on dislocations and grow almost instantly, with growth rates *reaching several kilometers per second. The precipitates take the form of platelets or needles* (Fig. 3.16).

Unlike phases formed by diffusion, the amount of new phase resulting from a military transformation does not change over time. It increases only when the quenching temperature decreases between two extremes, called the start and end temperatures of transformation, respectively.

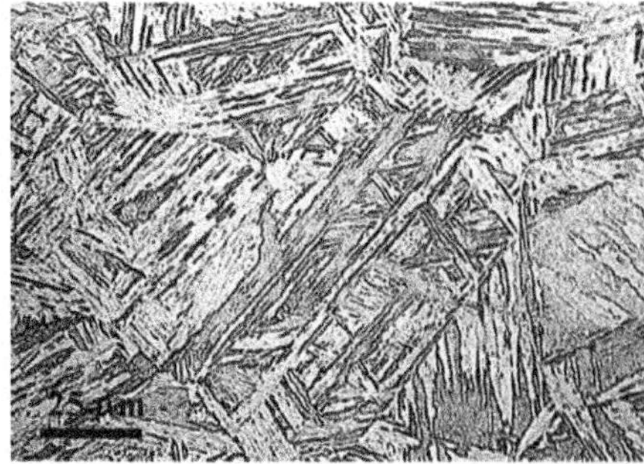

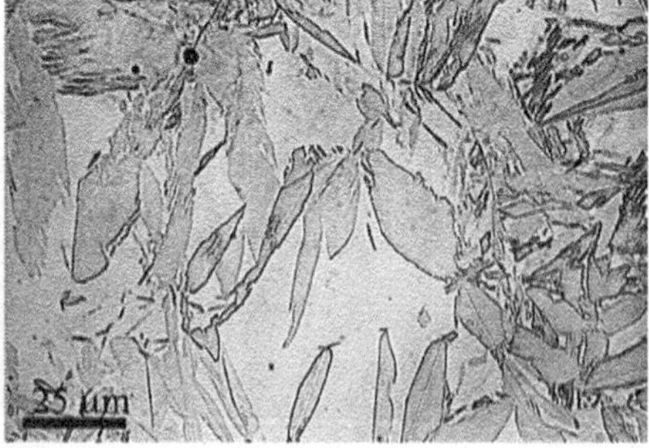

Fig. 3.16 Phase formed by quenching, without diffusion, of an iron and nickel alloy: the new phase (in gray) appears in the form of laths or needles depending on the nickel content, 28 and 30% respectively

Change of States in an Amorphous Material

The Solidification of an an amorphous material leads to a metastable solid state, passing through an intermediate region which strongly influences the material's properties.

Below the melting temperature of a material, the stable solid phase that forms abruptly corresponds to the crystalline state. However, in the case of amorphous materials, crystallization does not occur. An intermediate state between solid and liquid appears: *supercooled liquid* in the case of glasses and *rubbery state* in the case of polymers (Fig. 3.17). The glassy solid only exists below the glass transition temperature. In this case, we speak of a change of states rather than a phase transformation. The evolution of a property such as the specific volume (per unit weight) as a function of temperature allows us to define the domains of existence of these states.

The glass transition temperature is important: above it, the material can be softened and shaped. From a practical point of view, it should not be too high, but at the same time not too low so that accidental thermal fluctuations do not cause a sudden change in the material's properties during use. In the case of glasses, it depends on viscosity. The glass transition temperature decreases from that of pure silica (around 1,700 °C) with high viscosity to that of easily shaped glasses such as bottle glass and lead glass (around 700 °C). For amorphous polymers, the glass transition temperature is the temperature at which weak bonds begin to melt. It increases when

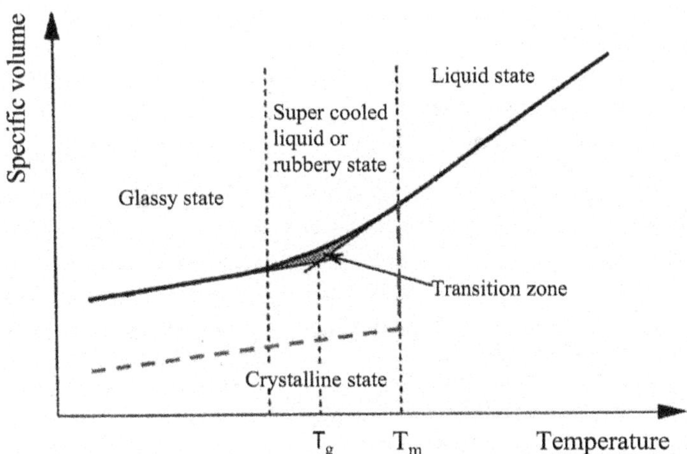

Fig. 3.17 Transition from the liquid state to the solid state for a crystalline material and an amorphous material. A transition zone appears for amorphous materials: supercooled liquid in the case of glasses and rubbery state in the case of polymers

the flexibility of the macromolecule chains decreases and when the average macromolecular mass (not all macromolecules have the same mass) increases. Finally, if the polymer is partially crystalline, only the amorphous fraction gives rise to the glass transition, the temperature of which approaches the melting point.

Dislocation Motion

Dislocation movement is the origin of a crystalline material's ability to permanently deform. Without dislocation, a crystal deforms like a rubber band and then breaks suddenly beyond a certain force.

Moving a heavy carpet placed on a floor by pulling it suddenly at one end requires a great deal of force. Conversely, it can be moved easily if a crease is created and then propagated from one end of the carpet to the other. To explain the low force required to move one part of a crystal relative to another by shearing, we use to the carpet analogy. The sudden displacement of the two parts of the crystal requires overcoming all the bond forces between atoms on either side of the shear plane and leads to theoretical values of the force required approximately 10,000 times greater than the actual force. The equivalent of a fold? It is a *dislocation* that moves step by step by *gliding* along a *slip plane* from one end of the crystal to the other. After a single passage, a step forms at each end of the crystal equal to the smallest distance between atoms. The resulting deformation is *plastic* (permanent), as opposed to *elastic* deformation, in which the crystal returns to its initial state as soon as the force ceases (Fig. 3.18).

Sliding does not involve any individual migration (no diffusion) of the atoms that move collectively. The passage of a large number of dislocations in the same plane and in different planes, and in different grains, results in significant deformations of a polycrystal. The glide velocity of dislocations varies by several orders of magnitude depending on the temperature, the stress applied to the material, the purity of the crystals, and their bond type (covalent, ionic, or metallic). It can reach 100 m per second (its theoretical limit is that of sound: 340 m/s), but most often it varies between 1 μm and 1 mm per second. It also depends greatly on the presence of other crystal defects that act as obstacles to glide.

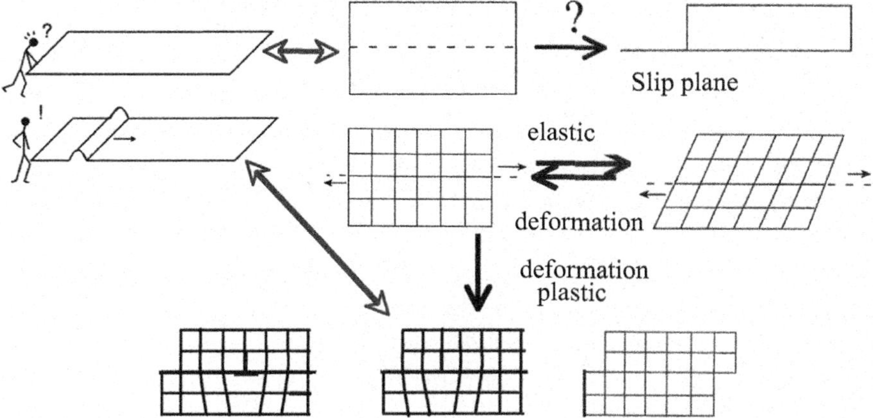

Fig. 3.18 Moving a carpet by propagating a fold is easily accomplished compared to the effort required to move it all at once. This is a good illustration of the basic process of plastic deformation, which proceeds by the sliding of a dislocation (⊥), analogous to a fold, from one end of a crystal to the other. A permanent step is created at each end of the sliding plane. Conversely, after elastic deformation, the crystal returns to its original shape as soon as the force ceases

At high temperatures, a dislocation can also move by *climbing*. This is a movement outside the slip plane accompanied by diffusion. Indeed, climbing corresponds to the extension or regression of an additional plane of atoms, and therefore to the addition or removal of vacancies depending on the direction of the dislocation's movement (Fig. 3.19).

But the presence of dislocations is not inevitable! Single crystals can be made in the form of very thin, *dislocation-free wires* called *whiskers*. These wires are very flexible; they respond elastically to bending. They are used, in elongated form like reeds, in luminous structures: they are transparent like glass because they are homogeneous, and they curve elegantly.

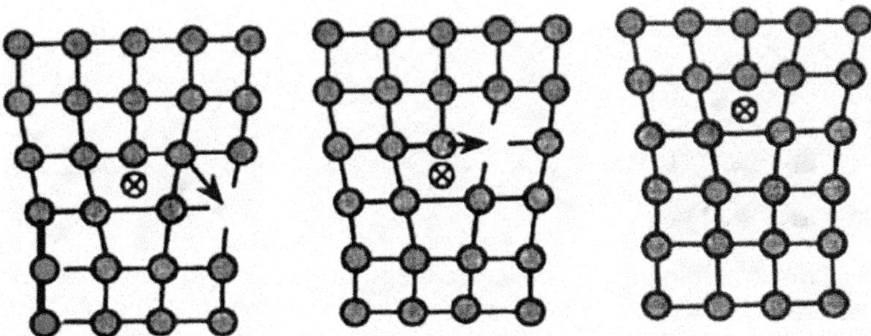

Fig. 3.19 Climbing of a dislocation by diffusion of vacancies. Between the initial state and the final state, the dislocation (⊗), has moved up one plane. The additional plane has lost atoms

Interactions Between Defects—Mobility Promotes Encounters

Dislocations move at high temperatures thanks to vacancies. The interaction of several dislocations leads to configurations that harden or, conversely, soften the material. Interfaces are obstacles to the movement of dislocations.

Mobile elements in the solid's space are capable of uniting, opposing, or annihilating each other, which leads to significant changes in the behavior of materials. As they move, dislocations encounter and interact to form configurations that harden or soften the material. Two dislocations combine if their total energy decreases during the combination, leading to a softening of the material. In the best case, the deformations associated with the dislocations annihilate, the energy is canceled, and the perfect crystal is obtained (Fig. 3.20).

A dislocation can be *anchored* or *pinned* at two fixed points; subjected to a force that forces it to slide, it then bends to form a semicircle and then wraps around the anchor points. The wrapping develops, and opposing strands along the dislocation line combine and annihilate, giving rise to a dislocation loop. The initial segment is recreated; the scenario can begin again like a *windmill*, provided the force continues: this is a *source of dislocations* (Fig. 3.21). These are not created *ex nihilo*, but multiply from existing dislocations in the material. Two dislocations sliding in different planes can block each other; they combine to form a stationary "lock." They can no longer move, and their contribution to deformation therefore ceases: the crystal hardens.

Dislocations moving in the same plane pile up on a grain boundary, which is an effective barrier to slip (Fig. 3.22); their passage from one crystal to another is significantly slowed or even prevented.

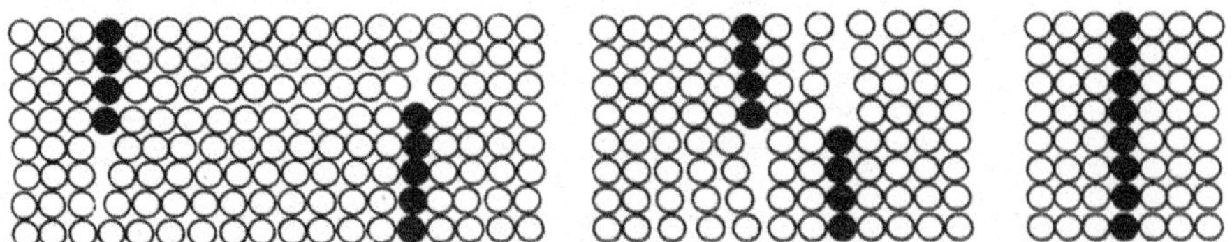

Fig. 3.20 Two-dimensional schematic representations showing how two dislocations in opposite directions move toward each other and annihilate. The movements are visualized by those of the additional planes (atoms in black). When the half-planes are facing each other, they form a complete plane and the dislocations no longer exist. The perfect crystal is restored

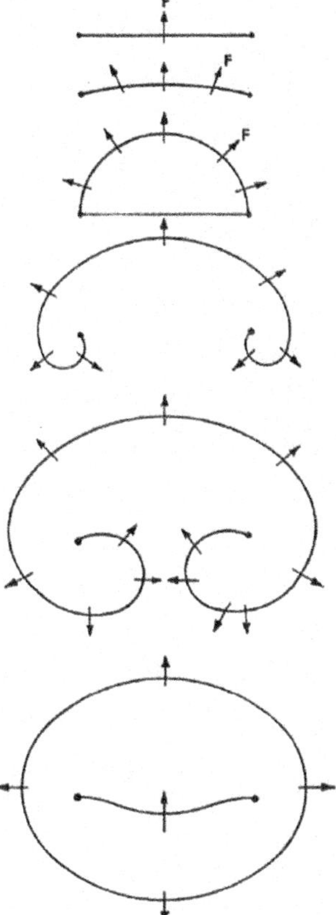

Fig. 3.21 Diagram of the operation of a dislocation source

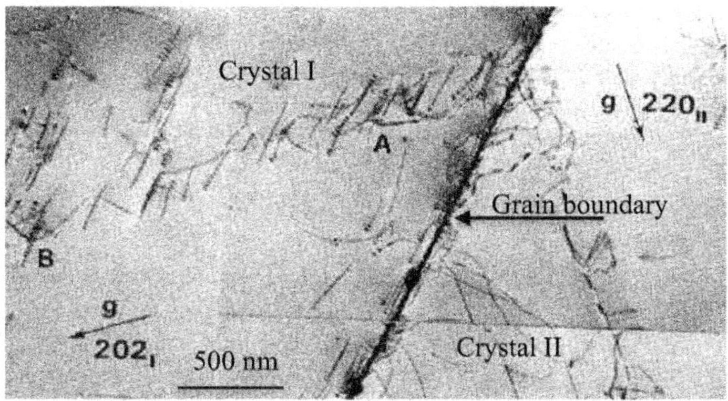

Fig. 3.22 Stack of dislocations (A) on a grain boundary in slightly deformed silicon

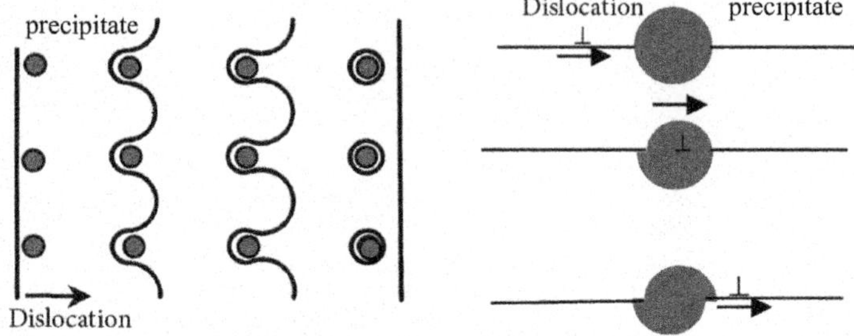

Fig. 3.23 Schematic representation showing: on the left, the bypassing of precipitates by a dislocation line; at the end of the bypassing, a dislocation loop surrounds each precipitate; on the right, the shearing of a precipitate by a dislocation

Finally, dislocations are also hindered in their movement by volume defects (precipitates, inclusions), especially when these are small and homogeneously dispersed throughout the crystal. They are forced to bypass or shear the precipitates (Fig. 3.23). These mechanisms are at the origin of *structural hardening*: to increase the hardness of an alloy, elements capable of forming small precipitates with the basic constituents are added.

All these mechanisms occur simultaneously during the plastic deformation of polycrystals, either synergistically or antagonistically; they must be considered to understand the behavior of the material.

4

Which Material? for What Use?

The discovery of the properties of materials and ways to shape them most likely originated in the decorative arts. The first creative impulse was an aesthetic experience, but using a material for useful purposes quickly required a better understanding of the relationship between its composition and its behavior under various stresses.

Two properties of materials are especially important: their mechanical strength and their resistance to corrosion. Even if an object is used for electrical or magnetic functions, the materials it is composed of must be shaped (malleability), not deform over time (fatigue resistance), and simultaneously not corrode or oxidize at high temperatures. A ceramic is neither corrodible nor oxidizable, but it is fragile! Conversely, some steels are very hard but easily corroded by seawater, causing major problems for reinforced concrete structures.

It should be remembered that the properties of a material are also controlled by the form and dimensions in which it is used: for example, solid gold is unalterable, while gold divided into extremely fine particles is highly reactive. Many recent applications of materials rely on their miniaturization.

Finally, considering the various properties of a material is not enough to guide its selection for a given application: potential toxicity, prohibitive or fluctuating costs, and difficulty in recycling are all factors that can overshadow its functional qualities.

L. Priester, *Materials: History, Science and Perspectives*,
https://doi.org/10.1007/978-3-032-15754-6_4

Electrical Conductor (Metal) or Insulator (Ceramic)

A material is a conductor or an insulator depending on whether or not it is crossed by a current when subjected to an electric field. Superconductivity* is the elimination of all electrical resistance.

In a solid *conductor* of electricity, current is the result of the movement, through the solid, of charged particles or *charge carriers*: electrons or ions*.

Thanks to the *electronic conductivity** of metals and alloys, we benefit from the "Electricity Fairy": lighting, heating, food preservation, and assistance with many industrial, agricultural, and household tasks. "The Electricity Fairy"? This is the title given by Raoul Dufy to his large painting created in 1937 for the Electricity Pavilion at the International Exhibition, now on display at the Musée d'Art Moderne de la Ville de Paris. Applications require conductive wires or cables, mainly copper, which connect devices to a current source. When a device is plugged in, the free electrons move in the metal conductor toward the positive pole of the generator, the current flows in the opposite direction, and the device can operate. Some polymers also conduct electricity, but only in the direction parallel to their chains. Conductivity in this direction, comparable to that of pure metals, can be 100,000 times greater than that in perpendicular directions.

Insulators are materials in which electrons are bound to nuclei (ionic or covalent bonds) and resist movement: glass, diamond, as well as most plastics and ceramics.

Resistivity characterizes a body's ability to oppose or not resist electric current. This quantity varies considerably: between the resistivity of copper, one of the best conductors, and that of Pyrex, a good insulator, the relationship is comparable to that between the diameter of an atom and the diameter of the sun.

The resistivity of metals decreases slowly as temperature decreases; for a strictly pure single crystal, it is theoretically zero at absolute zero. In reality, electrical resistance persists even at low temperatures; it is linked to crystalline defects in conductors, particularly their impurities, which slow the movement of electrons (Fig. 4.1). A good conductor must be as pure as possible, such as electrolytic copper (containing approximately 0.02% oxygen) used in power transmission lines and many domestic applications.

At the beginning of the twentieth century, it was discovered that the resistivity of some pure metals abruptly vanishes at a critical temperature of a few kelvins. This state, called *superconducting*, has sparked considerable research to achieve higher critical temperatures. A major milestone was

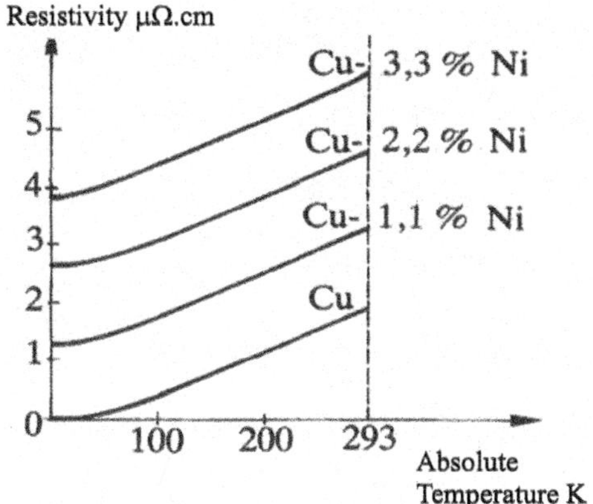

Fig. 4.1 Resistivity curves as a function of temperature for different purities of copper

reached in 1987 with mixed oxides of copper, barium, and yttrium, which reached a critical temperature of 95 K, then a year later, 133 K (–140 °C) with the addition of bismuth and calcium. Since then, this temperature has not been surpassed. The explanation for this phenomenon is complex; Well established for conventional superconductors (critical temperature close to absolute zero), it remains poorly understood for high-temperature superconductors (above –196 °C). The classical theory is based on the formation of electron pairs (known as Cooper pairs). Superconductivity occurs as long as the tendency toward electron ordering overcomes the effects of thermal agitation. This behavior is completely disconnected from the behavior that governs conductivity; it is therefore not surprising that a good conducting metal like copper does not exhibit superconductivity and, conversely, poorly conducting compounds exhibit a superconducting state.

The phenomenon of superconductivity has applications in several fields, including *medical imaging* by *nuclear magnetic resonance* (NMR) and *the storage of energy* in magnetic form in superconducting coils. It enables the construction of *elementary particle accelerators* that should lead to significant advances in fundamental physics and new applications in applied research. It also allows us to dream of the *magnetically levitated train* that glides a few centimeters above the track at over 500 km/h (Fig. 4.2). This is undoubtedly the most extraordinary application of superconductivity. Such trains exist on a real scale in prototype form. They have the advantage of very high speed combined with minimal track wear, but developing them and the infrastructure needed for their operation is not feasible today.

Fig. 4.2 Japan's Maglev train travels at a speed of 580 km/h, but it is not commercially available, and only a few privileged passengers have been transported at this speed!

From the Leclanché Cell to the Fuel Cell: Ionic Conductivity

Ionic crystals are the basis of the design of any electric battery. A fuel cell directly and continuously converts the chemical energy of a fuel into electrical energy, heat, and water. Its principle is simple, but its implementation is complex and expensive.

Ionic conductivity, another way of transmitting electricity, is observed in certain ionic crystals due to the presence of vacancies that allow an ion* to jump from one site to another, but it generally remains low. This conduction mode is used in *solid-state electrolyte batteries* (or *accumulators*) that convert the energy of a chemical reaction into electrical energy. It is widely used, from the zinc-anode Leclanché battery (1860) used in flashlights and toys to the *lithium batteries* used in watches, cameras, and pacemakers. This is a delicate technology; the cell must be very airtight because lithium reacts violently with water and nitrogen in the air, potentially causing explosions. To address this risk associated with the metallic form of lithium, *lithium-ions* or *lithium polymer* batteries have been developed since 1970, with their market emergence in 1991, for phones and laptops, as well as for electric vehicles.

A fuel cell is a solid-state electrolyte cell where at least one of the reactants is not contained within the cell, but supplied continuously an external source, with the reaction products (heat and water) being continuously removed. Among the various technological approaches depending on the material constituting the electrolyte, two are favored: the low-temperature polymer approach and the high-temperature ceramic approach. The *proton exchange membrane fuel cell* operates between 60 °C and 90 °C; positively charged hydrogen ions move from the anode to the cathode through a polymer electrolyte membrane that does not allow electrons to pass through.

A true electricity generator is obtained by stacking several "electrode/membrane" elements in series to form a battery. An interconnecting material allows the elements to be assembled together. The assembly is integrated into a complex system ensuring fluid management and fuel storage. These batteries are used in electric vehicles and for stationary or portable applications.

In a *solid oxide fuel cell*, which can be used at high temperatures (750–1050 °C), oxygen ions move from the cathode to the anode. Many ceramic materials serve as solid electrolytes, the most common being based on zirconium oxide. Fuel cells are highly efficient and hold promise as power supplies, but the materials (electrodes and electrolytes) need to be improved to reduce operating temperatures.

Electrical Conductor Under Certain Conditions: The Semiconductor

A semiconductor is a non-metallic, crystalline or amorphous material with covalent bonds, with intermediate electrical behavior: an insulator when pure at room temperature, it conducts electricity when heated or when impurities are added in homeopathic doses.

The properties of semiconductors stem from their electronic structure. In a solid, electrons have well-defined energies located in *energy bands* (*Fig. 4.3*). In their ground state, electrons are located in the *valence band*; this is generally full; its electrons cannot move because this requires empty energy levels. So-called "excited" electrons can, however, move into a higher energy band called the *conduction band*; they are then mobile and conduct electricity. This situation is easily achieved in some metals, because the valence and conduction bands overlap. In other metals, the valence band has empty states between which electrons move. In both cases, the metal is a good conductor. In other cases, there is an *energy gap* between the valence and conduction bands. To become mobile, an electron in the valence band must overcome the energy gap; the width of the band gap governs this possibility: if it is wide, no electron can cross it; we have an *insulator*. Conversely, if the band gap is narrow, electrons in the valence band can jump into the conduction band and move within this band; Their absence in the valence band creates *holes* or positive charges that are also mobile. The material exhibits so-called "*intrinsic*" *semi-conductivity*, because it involves only the electrons of the material itself (Fig. 4.3). The two types of charge *carriers*, electrons and holes, which ensure the propagation of electric current, are then in equal quantities.

Silicon and germanium are intrinsic semiconductors; they are ceramics with a covalent bond that is much weaker than that of an insulator like diamond.

So-called "*extrinsic*" *semi-conductivity* is obtained by doping the material with tiny amounts of an element that favors a higher number of electrons relative to the number of holes, or vice versa: we obtain an *N-type* semiconductor (N for negative) that donates electrons (Fig. 4.4), or, conversely, a *P-type* semiconductor (P for positive) that accepts electrons.

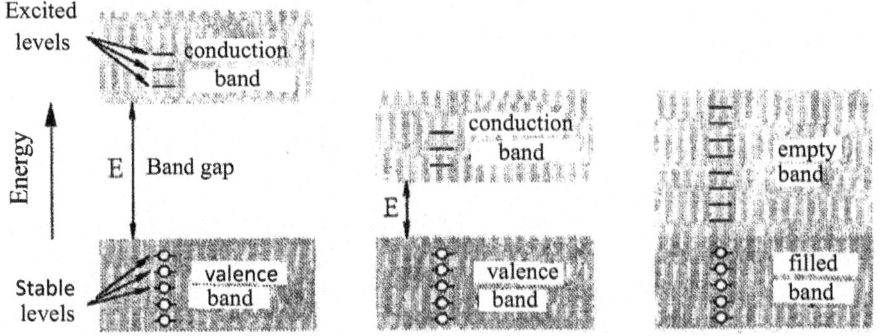

Fig. 4.3 Diagram of the energy bands for the three types of materials: insulator, semiconductor, and conductor

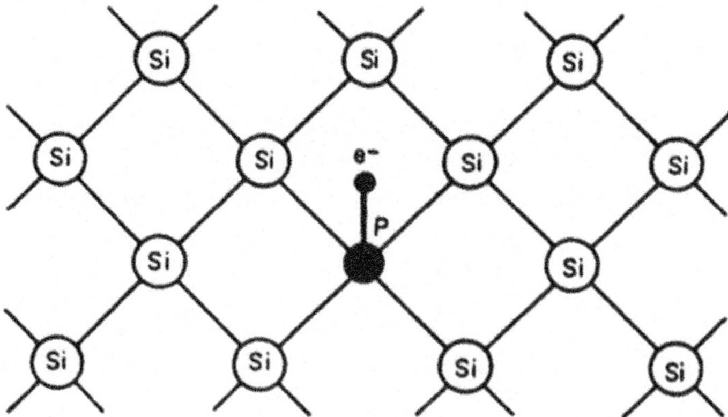

Fig. 4.4 Diagram of the bonds in an N-type semiconductor: silicon (Si) doped with phosphorus (P). The electrons of the phosphorus atom, except one, participate in covalent bonds with neighboring atoms. The additional electron (e–) is free and can move within the crystal

Intrinsic conductivity is weak, with the number of electrons released by an atom remaining very limited: it is one hundred million times lower for silicon around 750 °C than for copper at room temperature. It increases sharply with temperature due to the increase in the number of carriers: thus, it increases a 100-fold for silicon between 750 and 1500 °C.

Extrinsic conductivity is higher while remaining significantly lower than that of metals. It enabled the development of electronics, the simplest element of which is a *PN junction* or *diode* (Fig. 4.5), created by doping adjacent regions of a semiconductor with P and N dopants, respectively. If a positive voltage is applied to the P region, the majority positive carriers (holes) are repelled toward the junction. At the same time, the majority of negative carriers on the N side (electrons) are attracted toward the junction. This leads to an abundance of carriers at the junction, and electric current can flow through it. If the potential difference is reversed, the carriers move away from the junction, thus blocking the flow of current at that junction. A diode therefore only transmits current in one direction.

A third region can be doped to form a double N-P-N or P-N-P junction at the base of the *transistor* (transfer resistor), which is the fundamental active component in electronics, primarily used for current control and amplification. The invention of the transistor by Bell Laboratories in 1947 was decisive for the development of the electronic information society. Millions of transistors, covering a surface area of a few square millimeters, fit into an *integrated circuit* or *electronic chip*. In the 1970s, the first *microprocessors** contained a few thousand transistors; by 2007, they contained one or two billion (Fig. 4.6). During the same period, the diameter of the smallest wire connecting two microprocessor components has changed from a few micrometers to 0.045 µm.

Moore's Law proved surprisingly accurate until the beginning of the twenty-first century. In February 2016, Moore's Law was abandoned. Indeed,

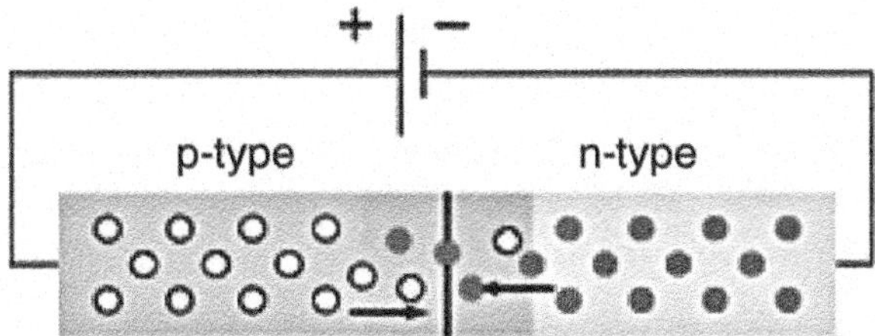

Fig. 4.5 PN diode formed, for example, from a single crystal of arsenic-doped silicon in the N-type region and aluminum in the other P-type region. The charge carriers, electrons and holes (solid and open circles, respectively), move along the crystal

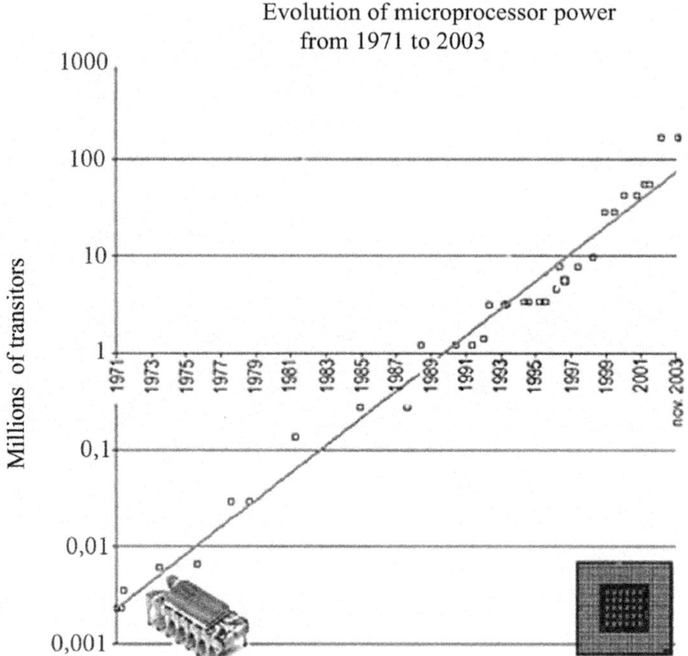

Fig. 4.6 Moore's Law predicts the evolution of the number of transistors in microprocessors. The dots indicate their actual numbers. At the bottom, Intel microprocessors from the 1970s and 2003

the industry is increasingly approaching the physical limits of microelectronics, where transistors will no longer consist of anything more than a few atoms and the insulation between them.

Thermal Conductor or Insulator: Copper and Glass

Temperature is an effect of the thermal agitation of atoms. If they vibrate in the same way, the temperature in the solid is uniform. Otherwise, the intense vibration of atoms is transmitted from hot regions to their cooler neighbors, which in turn heat up.

Thermal conduction corresponds to the transfer of heat through a material toward its coldest region, a transfer that is faster the higher the temperature gradient*. From an atomic perspective, thermal conduction results from two phenomena: the movement of free electrons and the thermal agitation of atoms, which is transmitted from one to another. The first phenomenon

Fig. 4.7 Insulating glazing in a building; insulating and solar protection for a veranda

is predominant in metals, the second in non-metals. Electrons only possess a small portion of heat energy (about 10%), but the transmission of this energy is very efficient because electrons can travel great distances. Like a good electrical conductor, a good thermal conductor is primarily a material in which electrons are free.

Thermal conductivity measures a material's ability to conduct heat. It is high in metallic materials, but considerably lower in other materials. Glass and many plastics conduct heat a hundred to a thousand times less than silver, one of the best thermal conductors. In reality, no solid is a perfect heat insulator. The quality of thermal insulators is mainly linked to their porous structure. In construction, glass wool, a mat of extremely fine glass fibers, or rock wool are used. However, the most common insulation system involves placing air between two walls; air is indeed an excellent insulator provided that no convection movement disrupts it. Double glazing is therefore made up of two sheets of glass sealed and separated by a sealed space containing air (Fig. 4.7).

Thermal insulation does not always go hand in hand with sound insulation. Thus, double glazing does not generally protect against all noise, especially traffic noise. Cork, a natural porous material, and certain elastomeric foams possess both the qualities of insulating against noise and heat.

The Mystery of the Magnet

All substances react to the application of a magnetic field. Magnetization is generally weak; however, a few rare materials are strongly magnetized and, alone, are of practical interest. They are called ferromagnetic even if they do not contain iron!

A solid is strongly influenced by a magnetic field if at least some of its atoms behave like small magnets. These are aligned within domains (Fig. 4.8); if these domains are arranged in all directions, the macroscopic field is zero. In the presence of an external magnetic field, the domains with orientation close to the field grow by migration of their walls. If the walls move easily, the material becomes easily magnetized, but only remains a magnet in the presence of an external field: it is a *soft magnetic material* used in transformer circuits. If the walls move with difficulty, the material becomes magnetized with difficulty, but demagnetizes slowly; it is a *hard-magnetic material*.

The response of a ferromagnetic material to a magnetic field is given by its magnetization intensity, which increases and then becomes constant beyond a certain field value. When this field is canceled, the magnetization persists: this is the *remanence* value. The magnetization is eliminated if the material is subjected to a field opposite to the previous one, known as a *coercive field*. The same phenomenon occurs if the field is increased in the opposite direction. These characteristics are plotted on a curve called the *hysteresis loop*; if the hysteresis is weak (little difference between the curves (2), the magnetic field is soft, and vice versa (Fig. 4.9).

A magnet is a *hard-magnetic material*; it is useful for picking up the contents of a box of needles scattered on the floor or for automatically closing a door. Very powerful permanent magnets, much more powerful than traditional magnets, are currently made from a neodymium/iron/boron alloy (NdFeB) (Fig. 4.10).

The phenomenon of *giant magnetoresistance*, discovered in 1988 and awarded the Nobel Prize in Physics in 2007, occurs in structures composed

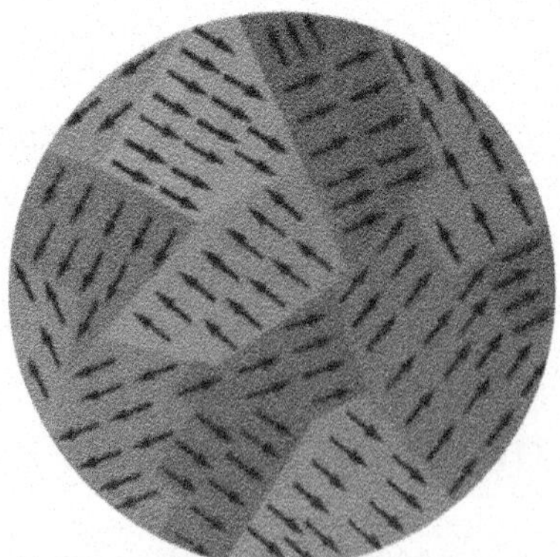

Fig. 4.8 Schematic representation of the domain structure of a ferromagnetic material. Each arrow corresponds to a small magnet

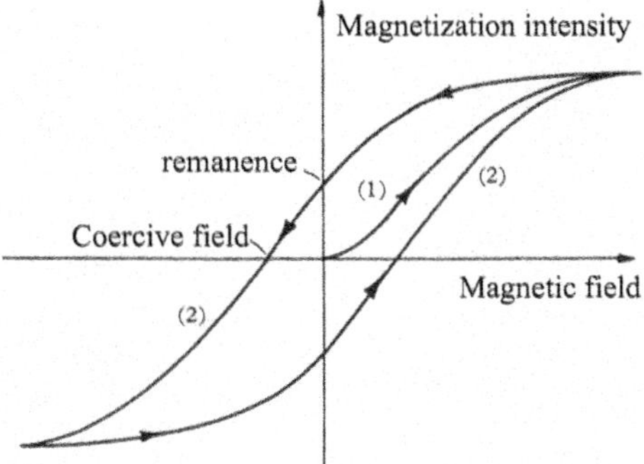

Fig. 4.9 First magnetization curve (1) and hysteresis cycle of a ferromagnetic material (2)

Fig. 4.10 Supermagnet with a nail cap. These neodymium/iron/boron alloy magnets are much more powerful than traditional magnets

of alternating ultrathin layers (a few atoms) of a ferromagnetic metal, such as iron, and a non-magnetic metal such as chromium.

When the iron layers are magnetized alternately, little or no current flows; however, a significant drop in the resistance of the multilayer occurs when the respective magnetizations of two adjacent ferromagnetic layers align under the application of an external magnetic field (Fig. 4.11).

This effect is the basis for the manufacture of highly sensitive read heads capable of reading information stored at very high density on hard drives: a small magnetic field, such as that generated by magnetic inscriptions on these drives, opens the door to electric current.

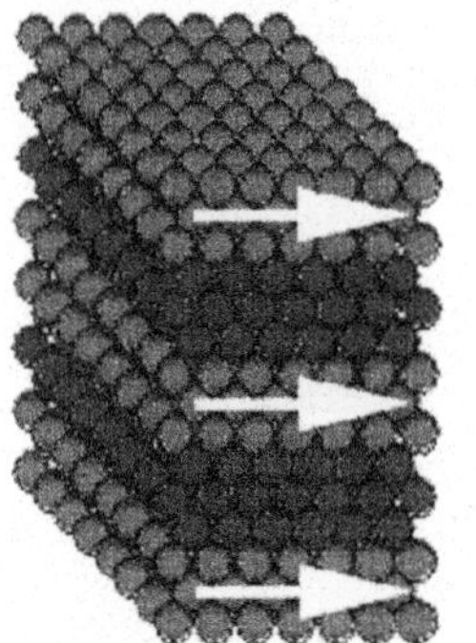
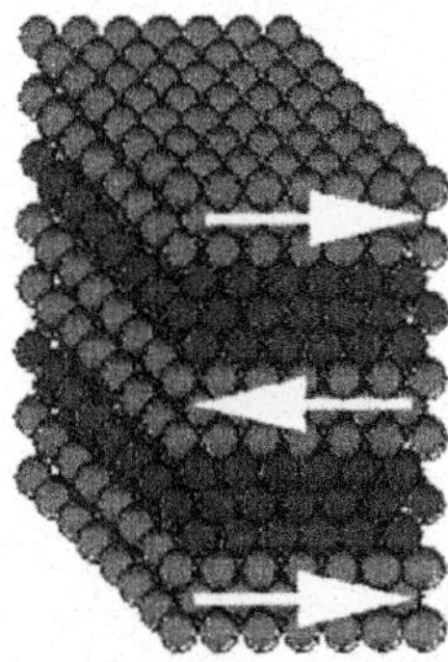

Fig. 4.11 Stack of layers of chromium (in black) and iron (in dark gray). When the iron layers are magnetized alternately (right), the electrical resistance is extremely high compared to the situation where the magnetizations are all in the same direction (left), as is the case in the presence of a magnetic field

Reflecting and/or Transmitting Light

Solids absorb light to varying degrees, reflect it, and/or transmit it; some are colored, others white or transparent. Liquid crystals hold a special place in the world of optics: their response to light varies under the effect of a very weak electric field. This phenomenon is the origin of liquid crystal displays that have invaded the world of computing and television.

Electrons in metals absorb light in a surface layer less than 100 nm thick. The acquired light energy causes them to move into the conduction band. When they return to the valence band, they emit radiation by *scattering* or *reflection.* These shifts do not occur in *insulators*, meaning that light penetrates the material where it can undergo *absorption* (energy dissipated as heat), simultaneous *refraction*, and, possibly, *transmission* (Fig. 4.12).

Color and *transparency* result from these phenomena. A metal is generally white because all the light is reflected. It appears colored if one (or more) color of the white light is selectively absorbed, then it takes on the *opposite* (*complementary*) color; for example, copper absorbs blue light, returns light devoid of blue, and therefore appears red. Insulators, such as diamond and very pure polymers, are transparent. But if the light is slowed by impurities or second-phase islands, the material becomes tinted or opaque depending on whether the absorption is partial or total.

The *refractive index* is the ratio between the speed of light in a vacuum (equal to 1) and this speed in a transparent material, which is always less than unity. Diamond (crystalline carbon) has a high refractive index, followed by lead glass (or crystal), and then polymers. If the light originates from the

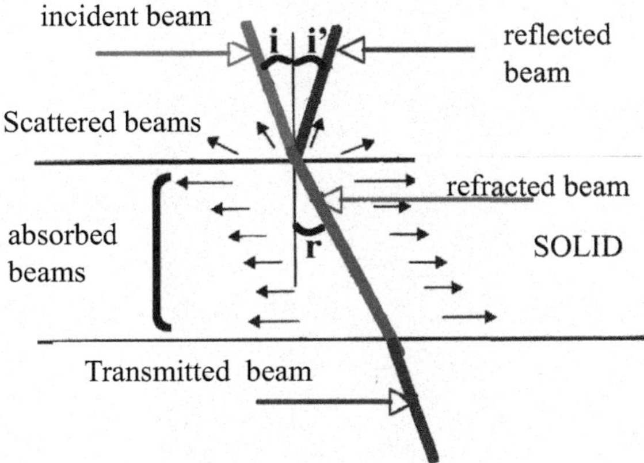

Fig. 4.12 Diagram showing the different interactions of a monochromatic ray falling on a plate of a material. The angle of incidence (i) is equal to the angle of reflection (i'). The angle of refraction in the material (r) is always smaller than the angle of incidence

material, there is a critical angle of incidence beyond which refraction (in air) is no longer possible. An *internal reflection* phenomenon occurs, which is used to channel the light and transmit it with minimal energy loss (Fig. 4.13). This is the advantage of *optical fiber* (Fig. 4.14) with a quartz core and polymer periphery, a field almost exclusively used in telecommunications.

Combined with the polarization* of light, the application of an electric field to a liquid crystal (see Chapter II) is the basis of *liquid crystal* display technology, commonly known as LCD (*liquid crystal display*). A thin film of

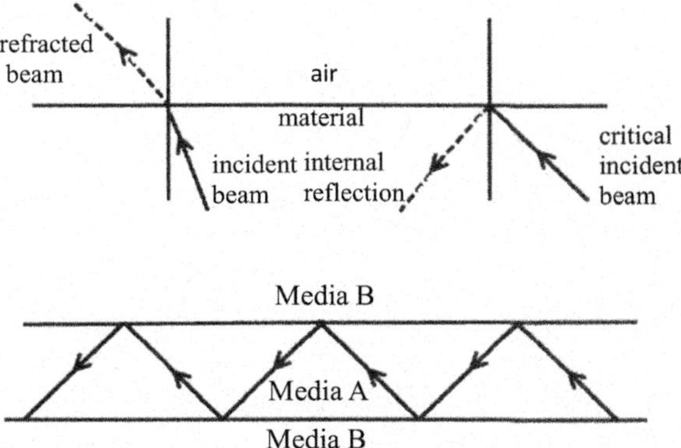

Fig. 4.13 Principles of internal reflection and of an optical fiber: the core is made of a material with a higher refractive index than the periphery. The ray entering one end of the fiber undergoes multiple internal reflections and propagates to the other end without loss, following a zigzag path

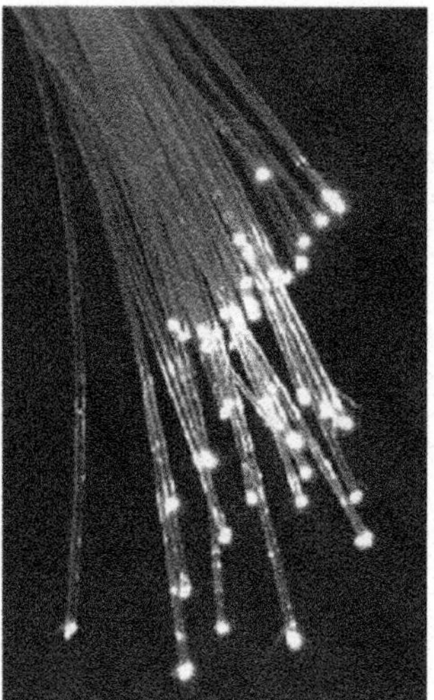

Fig. 4.14 Optical fibers can be presented in the form of a bouquet

nematic crystals, sandwiched between two glass plates to which two optical filters or *polarizers** are attached, transmits or blocks light depending on the orientation of its molecules. The assembly can constitute a microcell in a display whose lighting can be electrically controlled. One or three microcells per pixel are required depending on whether the image is monochrome or color. Current displays use the *twisted nematic effect*: the molecules are arranged in a helix within the liquid crystal. Light, oriented parallel to the molecule axis, after passing through the first polarizer propagates within the cell following a helical path imposed by the molecular structure. The second polarizer (or "analyzer"), rotated 90° relative to the first, receives light also rotated 90° and allows its transmission: the cell is turned on. The alignment of the molecules, induced by an electrical voltage, causes the light to propagate without deflection and is stopped by the analyzer: the cell is turned off (Fig. 4.15).

Liquid crystal displays provide display on watch faces and cell phone screens. They are used in flat screens for computers and televisions.

Cholesteric liquid crystals are widely used for their selective light reflection properties. For example, an object viewed at a 90° angle of incidence appears red, then changes to yellow, green, and then blue at increasingly oblique angles of incidence. This phenomenon, which gives rise to the iridescent colors of certain insect shells, is used in paintings and logos whose color changes with temperature or the angle of observation (banknotes).

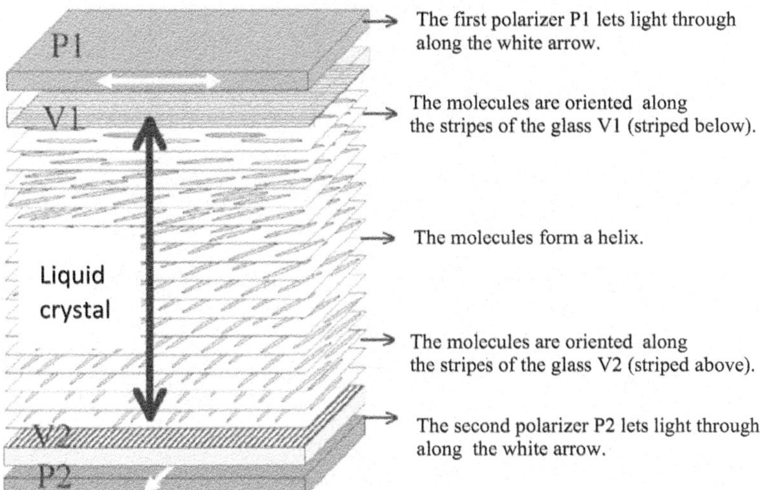

Fig. 4.15 Principle of a liquid crystal display cell using the twisted nematic effect

Resisting a Force: The Various Modes of Deformation

Under the effect of a force, a material resists or deforms. This mechanical behavior is a major criterion for choosing a material, regardless of the scale of its use: mass-produced sheet metal or microelectronic components.

A material's ability to deform is one of the first properties that humans implicitly exploited to manufacture objects. Mechanical behaviors vary depending on the class of materials: ductile metals, fragile ceramics and glasses, and elastic polymers. However, the differences between materials within the same class and between various microstructures (annealed, cold worked) of the same material are also significant. The conditions of deformation: temperature, stress mode (tension, compression, bending, torsion, etc.), and environment (dry or humid air, marine environment, high-temperature gas, etc.) also strongly influence the reaction of a solid subjected to a force. This complexity requires the definition of a test to assess the strength of a material: the *tensile test*, in which a specimen (rod or bar) (Fig. 4.16) is subjected to an increasing force and its elongation is measured. The objective is to establish a relationship between the applied stress (force per unit cross section of the specimen) and the deformation (elongation per unit length).

The specimen initially deforms *elastically*, meaning it returns to its initial length when the force ceases. Beyond the *elastic limit*, it acquires permanent deformation: this is the stage of *plastic deformation* that continues until fracture (Fig. 4.17). A metallic single crystal deforms easily at the very beginning

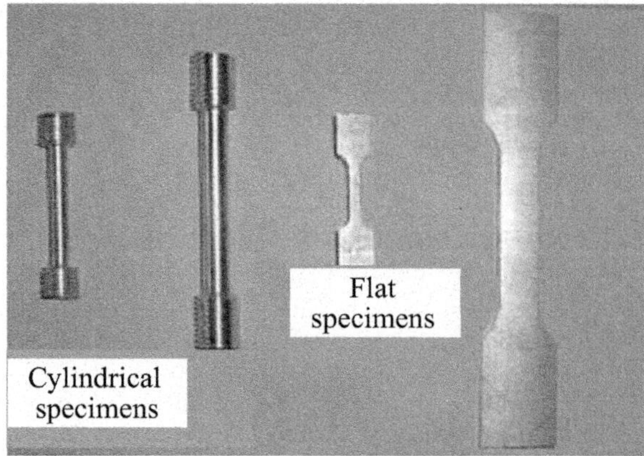

Fig. 4.16 Examples of tensile specimens made of aluminum alloy: their ends are fixed in jaws and stretched. The tests are performed in a tensile testing machine

of the plastic region (plateau on the tensile curve). On the other hand, at low-temperature, ceramic breaks immediately after its elastic limit (Fig. 4.18). These behaviors can be explained by the movements of dislocations and their interactions (see Chapter 3). A single crystal deforms easily because the dislocations slide in parallel planes without intersecting, and therefore without hindering each other, and no grain boundaries block them. In contrast, the deformation of each grain in a polycrystal is hindered by those of its neighbors and requires a constant increase in the applied force. In a ceramic where covalent or ionic bonds are strong, a high stress, close to the breaking stress, is required to cause dislocations to slide. If the temperature increases sufficiently, the bonds weaken and the ceramic becomes plastic.

Glass deforms above its glass transition temperature by *viscous flow*: covalent bonds break and re-form. An amorphous material often exhibits *viscoelastic behavior*, with both viscous and elastic behaviors coexisting. Some polymers deform *plastically* without the *intervention of dislocations*, even if these are present in the crystallized regions. Deformation essentially occurs through the reorientation of amorphous macromolecule chains and crystalline lamellae. The chains disentangle themselves more easily when they are shorter, their network is less dense, and the temperature is higher (Fig. 4.19).

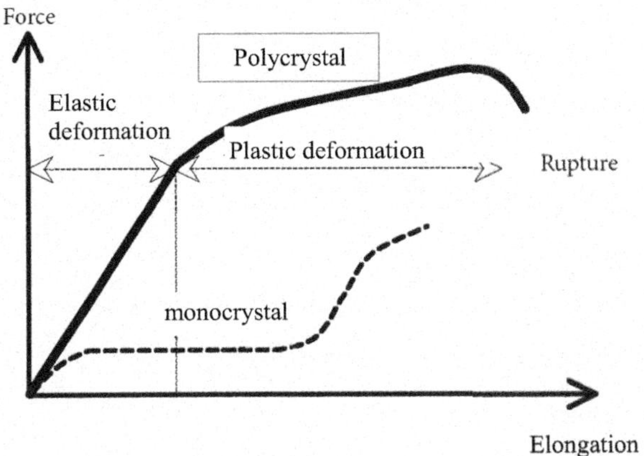

Fig. 4.17 Aspects of tensile curves relating force to elongation for the same monocrystalline and polycrystalline metal

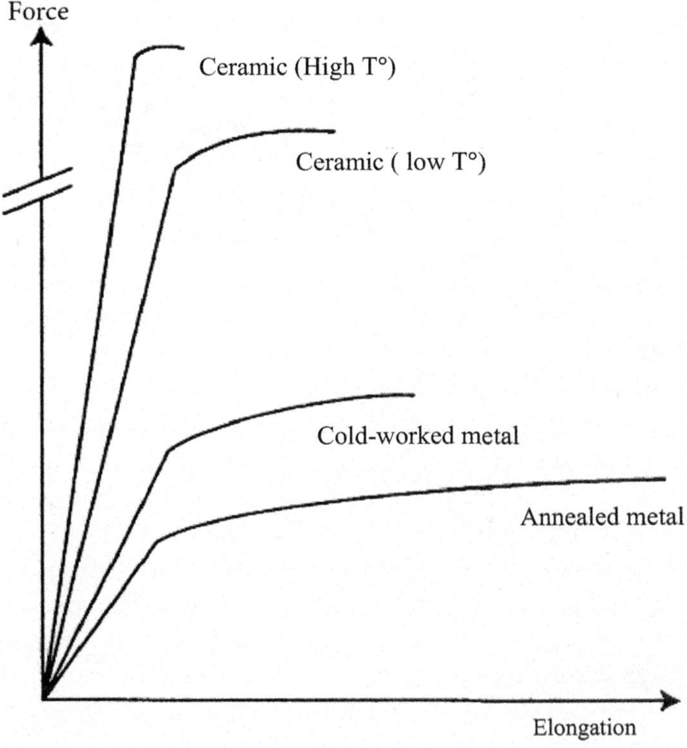

Fig. 4.18 Aspects of tensile curves relating force to elongation for the same annealed or work-hardened metal and for a ceramic at high and low temperatures (the strength in the latter case is significantly higher)

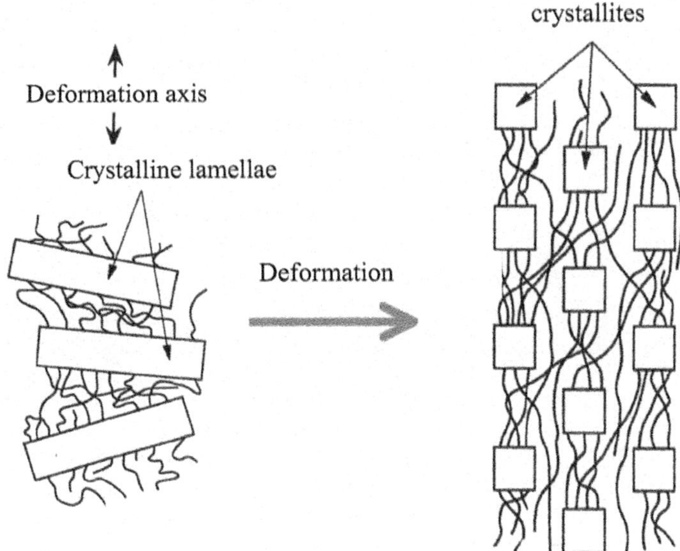

Fig. 4.19 Structural reorganization in a semi-crystalline polymer due to severe deformation. The scales are not true to form: the extension after deformation is much greater than that shown

Withstanding a Force How Long?

Regardless of the precautions taken when using a material, it wears over time and eventually breaks. Failure can occur normally, predictable by creep or fatigue tests, but can also occur catastrophically.

In a tensile test, it is assumed that if the force is no longer increased from a value lower than that which causes failure, the deformation of the material ceases, and its microstructure no longer changes. This assumption, valid at low temperatures, is no longer valid when the temperature increases.

Creep reflects the ability of a material to deform over time under the effect of a constant, relatively low stress. Deformation is only significant if the stress is applied at a temperature greater than about one-third of the absolute melting temperature (T_m). This is the temperature range for many applications (aeronautics, energy, etc.) in which materials are subjected to hot stresses. In *creep testing*, a stress, generally tensile or compressive, is applied to a specimen and its length variation is measured over time (Fig. 4.20). The strain rate quickly becomes constant over a wide range of stresses and temperatures. This stage accounts for 90% of the specimen's lifespan and can last thousands of hours.

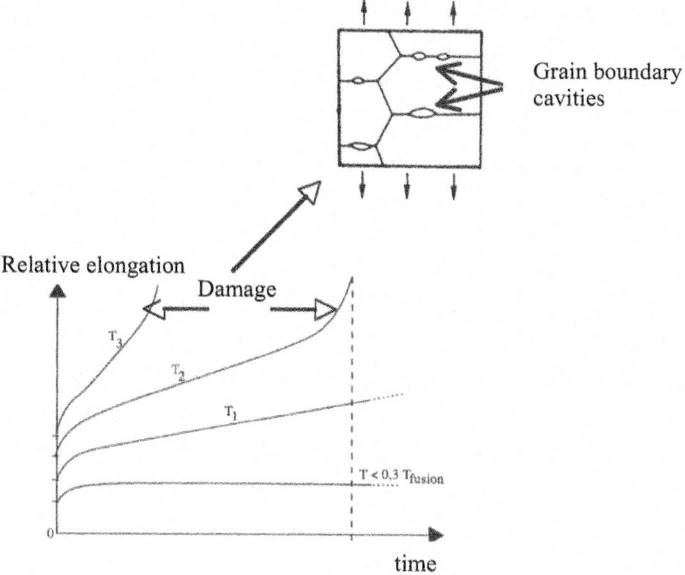

Fig. 4.20 Creep curves showing the various stages of creep at different temperatures (T3 > T2 > T1 > 0.3TF). The large linear portion corresponds to a constant rate. Damage occurs more quickly as the temperature (or applied stress) increases

Materials with nanoscale grains have the ability to deform under high temperatures (several hundred percent). This *superplasticity* is mainly due to the sliding of the grains relative to each other.

The lifespan of a material subjected to constant stress is estimated from a creep test at a temperature significantly higher than its intended use. An empirical relationship, based on the principle of time–temperature equivalence, allows for the deduction of the time to failure under operating conditions. However, the part must be monitored well before this time. Despite all precautions, a part can fail catastrophically. This *sudden rupture* is linked to the rapid propagation, at the speed of sound, of pre-existing cracks that propagate either because they have reached a critical size or because their ends are subjected to a critical stress. The stress at the crack tip is indeed much greater than that imposed on the entire material: there is a stress concentration. One of the advantages of composites is that the fibers stop crack propagation (Fig. 4.21).

A material can break under very low stress if it is applied cyclically. This situation often occurs in practice: aircraft landing gear, bridges, axles, engine parts, and our bones are subjected to stress alternating with periods of rest or fluctuations in the applied force over time. An overused staple eventually breaks. The formation and then slow growth of cracks reflect material fatigue.

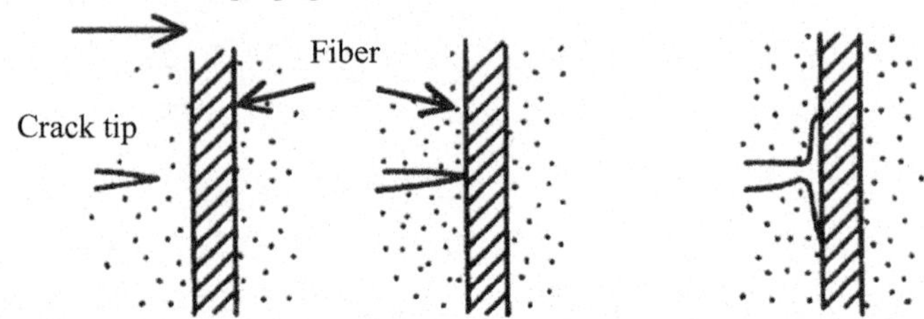

Fig. 4.21 Diagram of crack arrest by a fiber in a composite

To study it, a pre-notched specimen is often used, which is subjected to "tension/compression" or "alternating torsion" cycles of various shapes, and the number of cycles before failure is recorded (Fig. 4.22).

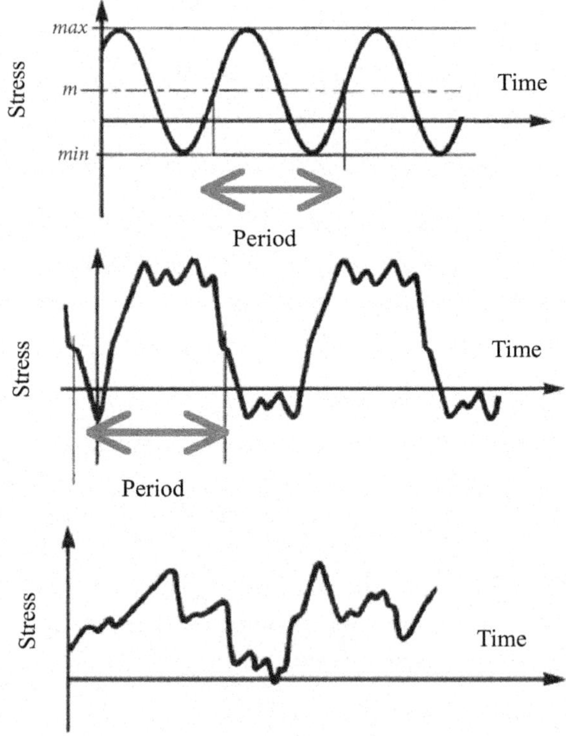

Fig. 4.22 Different types of cycles are imposed on a fatigue specimen: sinusoidal, periodic, and random, depending on the stress mode to which the material is subjected in practice

Surface Treatments and Coatings: Often a Necessity

The surface, more than the core of a part, is subject to mechanical erosion and/or corrosion by the environment. Its appearance must be aesthetically pleasing. Meeting these specific requirements requires "surface treatments," or the deposition of a material different from the substrate or coating.

A material's surface can be protected either by modifying its microstructure or chemistry in a layer of limited thickness close to the surface, or by applying a *coating*.

Surface hardening hardens the periphery of a part, particularly steel, relative to its center. It can be achieved by melting the part at the surface using a laser beam. A crack-resistant surface layer is formed by *carburizing* or *nitriding* (diffusion of carbon or nitrogen to the surface) or by other thermochemical treatments such as *carbonitriding* or *boriding*. *Anodic oxidation, sulfurization* or *phosphating* also contributes to the formation of protective surface layers against corrosion. In addition, the oxide protects against scratches and can give a beautiful appearance to a metal such as aluminum (Fig. 4.23) or titanium.

A hard, friction- and corrosion-resistant surface alloy can be formed on a metal part by ion *implantation* (nitrogen, carbon, zirconium, and platinum): the accelerated ions penetrate a layer of the material about a micrometer thick and transform it into a protective phase. Finally, mechanical techniques can be used to harden surfaces by work hardening: *shot peening* (blasting steel,

Fig. 4.23 Examples of anodized aluminum profiles used to create structures with very high strength

Fig. 4.24 A park gate exposed to the air or a boat hull in contact with seawater, a very dangerous corroding agent, is necessarily covered with a coating

glass, or ceramic balls), *hammering*, etc. Surface treatments, whether chemical, mechanical, or a combination of both, are applied before any use of steels to make them corrosion-resistant. Steels would be nothing without their surface treatments: in our environment, raw steel does not exist.

A *coating* (Fig. 4.24) is a material that is deposited in a thin layer on a substrate of a different nature. It generally provides good surface resistance to corrosion, high-temperature oxidation, wear, and crack initiation. It can also be deposited for aesthetic purposes. Some extremely thin coatings, consisting of a few atomic layers, are obtained by *electrolysis*, *physical deposition*, or *chemical vapor deposition.* The species to be deposited are either evaporated or sprayed and condense on the substrate (physical deposition), or they result from chemical reactions between gases at high temperatures (chemical deposition). Aesthetic protective coatings are obtained by *depositing particles*, particularly glass or ceramic, on metal or ceramic substrates: this is called *enameling*. Finally, thicker deposits include paints, plating (bonded layers or fixed by thermocompression, etc.), and layers formed by immersing the part to be protected in a bath of molten metal: zinc for galvanizing, tin for tinning, etc. Like the materials themselves, coatings have a limited lifespan;

they are attacked by humidity, especially near the sea, by a polluted environment, or by bacteria. They therefore require constant monitoring to prevent the garden gate from becoming covered in rust and the boat hull from being pierced by chlorine ions.

Conclusion

Knowledge of materials, from the performance required to produce a device to the most detailed descriptions of their composition, is constantly progressing thanks to the development of mechanical testing machines, electrical and magnetic measuring devices, optical-electronic instruments, and a whole range of computer simulations. Between the mid-twentieth century and the beginning of the twenty-first century, the spatial resolution (the smallest observable distance between two points) of a *transmission electron microscope* fell from around ten nanometers to less than 0.1 nm. We can now actually see a defect in a crystal through the movements of the atoms it carries around. The *simulations* cover a wide range of scales, from the macroscopic domain, for example, the distribution of forces experienced by a part (finite element simulation) to the nanoscale, with the arrangement and nature of atoms in a structure and in the vicinity of a defect (molecular dynamics simulation). At the mesoscopic scale, *which describes the intermediate length scales between atomic physics and those of the macroscopic world*, we can simulate the collective behavior of defects.

Experiments, simulations, and theoretical reflections have enabled us to better understand why a particular material resists deformation, corrosion, and the propagation of an electric current better than another. From then on, we were able to develop existing materials or design new ones. A major leap forward was made when humans realized the importance of the morphology of a material or an assembly of materials. In this discovery, man was aided by a new and more astute perspective on nature: a reed, a sponge, or a bark are remarkable materials. From these natural models, several ranges of composites and hybrid materials of increasing richness and complexity have been

L. Priester, *Materials: History, Science and Perspectives*,
https://doi.org/10.1007/978-3-032-15754-6

developed. Another aspect of this "mimicry" is miniaturization (nanometric dimensions), which confers surprising properties to a material, completely different from those it possesses in its bulk. The trend toward miniaturization extends from classic metals such as gold, iron, or copper to the most sophisticated ceramics and polymers.

Our foray into the world of materials also revealed that beneath the scientific term *microstructure* lies a great beauty in the forms that metals, ceramics, and polymers take on in their innermost being. These forms are often reminiscent of those observed in nature and, like them, can inspire artists and artisans: abstract painters, sculptors, fabric designers, and more.

Although far from exhaustive, the information provided in this book on materials opens up a better understanding of the objects we use personally and of all those that enter into today's life, whether in transportation, sports, or means of communication. Every day, every human being uses several materials: understanding them allows us to make choices. Which material should be used for future technologies? Which material should be used to avoid damaging nature? How can we recycle materials? How can we overcome the shortage of a material in a deregulated global trade? The knowledge acquired in this book should help the reader analyze society's responses to all these questions that affect the environment to a greater or lesser extent.

Glossary

Anodization: Superficial oxidation of a metal by electrolysis to improve its corrosion resistance or change its color.

Astrolabe: Ancient instrument used to obtain a simplified plane representation of the sky at a given date; the modern astrolabe allows one to observe the instant when a star reaches a specific height.

Autoradiography: Image produced directly on an emulsion or photographic film by the radiation of a radioactive substance.

Carbide: Combination of carbon with another simple substance. Silicon carbide and tungsten carbide are very hard.

Conductivity: Property of substances that transmit heat or electricity; also physical quantity that measures the capacity for conduction (the action of transmitting). The inverse of conductivity is resistivity.

Electrode: The end of each of the conductors attached to the poles of an electric current generator. The anode is the negative electrode during discharge and positive during charge; the opposite is true for the cathode.

Electrolyte: A substance that dissociates into ions under the effect of an electric current.

Ellipsoid: A volume whose cross sections are ellipses or circles.

Gradient: The rate of change of a quantity in a given direction.

Hydrophobic/hydrophilic: Insoluble/soluble in water.

Hydroxide: The name for compounds containing at least one oxygen/hydrogen (OH) group; for example, sodium hydroxide (NaOH) or aluminum hydroxide (Al(OH)3).

L. Priester, *Materials: History, Science and Perspectives*,
https://doi.org/10.1007/978-3-032-15754-6

Ion: An atom or molecule that has lost or gained one or more electrons. Ions therefore carry a charge, either negative (anion) or positive (cation).

Isotropic: A body that possesses the same properties in all directions, as opposed to an anisotropic body whose properties vary depending on the direction.

Kaolin: A form of clay (hydrated alumina silicate) used to make vitrified ceramics: porcelain, earthenware, tiles, bricks, etc.

Slag: A metallurgical by-product composed mainly of silicates and formed during melting to bind impurities.

Weak bond: Electrostatic attraction between electrical charges of opposite signs, with a significantly weaker intensity than covalent, ionic, or metallic bonds.

Ingot: A parallelepiped-shaped mass of metal or alloy obtained by pouring molten metal into a mold called an ingot mold.

Mesoscopic: An intermediate scale between microscopic (literally observable with a microscope) and macroscopic (observable with the naked eye). In metallurgy, microscopic observations are used for those involving a single crystal or a limited number of crystals. The term mesoscopic is reserved to describe the organization of a much larger number of crystals, which nevertheless remains smaller than that found in a macroscopic object.

Metastable: Refers to a system that is not stable, but appears so because it evolves at an extremely slow speed.

Microprocessor: A microprocessor is a processor (i.e., the component of a computer that performs the arithmetic and logic operations contained in programs) whose components are sufficiently miniaturized to fit on a single integrated circuit.

Nitride: A combination of nitrogen and another simple substance. Silicon nitride is very hard, like its carbide.

Golden ratio: A number equal to $(1 + \sqrt{5})/2$ (or ≈ 1.618), corresponding to a proportion considered divine. Throughout history, painters and architects have used the golden ratio to organize their spaces, resulting in particularly harmonious works.

Obsidian: Natural volcanic glass used in the manufacture of prehistoric tools.

Oxidation: A reaction in which an atom or ion loses electrons.

Oxide: A substance resulting from the combination of oxygen with another element (for example, alumina is aluminum oxide).

Phase: A region of a material that has homogeneous physical and chemical properties. Although of the same chemical nature, water and ice are different phases. Water plus ice constitutes a two-phase mixture.

Photovoltaic: A photovoltaic cell is an electronic component that, when exposed to light, generates an electrical voltage.

Polarizer: An optical device used to polarize light, i.e., to select the radiation that is parallel to its axis from among the rays that constitute white light.

Precipitation: The appearance of a second phase in the form of particles (visible under a microscope or to the naked eye) in a previously single-phase material.

Raku: A rapid ceramic firing technique pioneered in Japan and used worldwide. The incandescent pieces are smoked, dipped in water, burned, and left exposed to the air.

Annealing: Heating a metallurgical product or glass to a temperature sufficient to ensure its equilibrium or stress relief; it is followed by slow cooling.

Reduction: Reaction in which an atom or ion gains electrons.

Resolution: Smallest perceptible quantity of the quantity to be measured (spatial resolution, weight resolution, etc.).

Segregation: Local overconcentration of an element in a region of the material opposite its homogeneous concentration throughout. The element is not involved in a compound, otherwise it is called precipitation.

Silicate: A mineral found in the composition of many rocks, consisting of tetrahedra with a silicon atom in the center and four oxygen atoms at the vertices.

Tomographic atom probe: A device that allows a three-dimensional image of a material's composition to be obtained; this image is reconstructed by a computer, layer by layer, and atom by atom, with a spatial resolution close to the atomic scale. The lateral positions of the different ions on the surface of the sample are provided by the positions of their impacts on a 2D detector, with the lightest ions arriving first.

Sulfide: A combination of sulfur and an element; many sulfides are toxic, such as hydrogen sulfide or sulfur dioxide.

Superalloy: A complex alloy with very good mechanical and chemical resistance (oxidation and corrosion) at high temperatures.

Tetrahedron: A solid in space with four faces, six edges, and four vertices.

Credits

Unless otherwise stated, the diagrams presented in this book were created by the author.

t : at the top of the page

b : at the bottom of the page

Chapter 1

p. 9: RMN/Ch. Jean; p. 10: Peter Newark; p. 11: Prof. Ralph Solecki; p. 12: Prof. Joseph R. Caldwell; p. 15: Smithsonian Institution; p. 16: Personal collection; p. 17: Lee Boltin/Smithsonian Institution; p. 19: Musée des Invalides; p. 20: Cité des Sciences; p. 22: Snecma Motors; p. 26: Michael Reeve; p. 29: Musée de Poitiers/Ch. Vignaud; p. 34: Guillemette L'Hoir; p. 36: Chris Jordan/CCBY 2-0; p. 41b: Jerome Chatin/CNRS Photo library.

Plates chapter 1:

Plate 1: Dapper Museum, Paris; Plate 2: Museum Petit Palais, Paris; Plate 3: Archeological museum of Ankara; Plate 4: Nancy school museum; Plate 5: Museum of Chatillon-sur-Seine; Plate 6: Personal Image; Plate 7: Bernard Gasté/Les Chevaliers Carnutes; Plate 8: Gordon Taylor, Stony Brook University.

Chapter 2

p. 47: R. Zerbst/Taschen; p. 48: W.D. Kingery; p. 49: SFV; p. 50: N. Boriand; p. 55: L. Priester; p. 56t: S. Lartigues; p. 56b: S. Faisal; p. 57: L. Priester; p. 59: H. Gabrisch; p. 63t: after M. F. Ashby and D. R. H. Jones; p. 63b: A. Guinier; p. 64: A. Guinier; p. 71t: French Steel Federation; p. 71b: after M.

L. Priester, *Materials: History, Science and Perspectives*,
https://doi.org/10.1007/978-3-032-15754-6

F. Ashby and D. R. H. Jones; p. 72t: E.V. Grove; p. 72b: N. Boriand; p. 73t: after M. F. Ashby and D. R. H. Jones; p. 73b: Solid Reactivity Laboratory/ CNRS/Univ. of Burgundy; p. 74: S. Lartigue; p. 75: after M. F. Ashby and D. R. H. Jones; p. 76: L. Priester; p. 77: Y. Bréchet; p. 78: National Museum of Natural History.

Plates Chapter 2

Plates 1 and 2: Thierry Baudin (Paris-Saclay University - France)

Chapter 3

p. 80 and 81t: J. Bénard et al.; p. 81b: A. Guinier; p. 83t: V. Alonzo, p. 83b: I.L.W. Wagner and T.R. Mager; p. 84t and 84b: J. Bénard et al.; p. 85: T. Ogura et al.; p. 86b: M. Yan et al.; p. 88b and 89t: after D.A. Porter and K.E. Easterling; p. 90t: U. Dahmen; p. 90b: L. Priester; p. 93b: J. Douin; p. 94: B. Chalmers; p. 95t: J. Philibert; p. 95b: M. Martinez-Hernandez et al.

Plates Chapter 3

Plate 1: Didier Blavette (University of Rouen - France)

Chapter 4

p. 99: Guinier/Jullien; p. 100: Fysisk institute, University of Oslo; p. 102t: C. Oudet; p. 102b: B. Chalmers; p. 103: S-Kei; p. 104: according to J.B. Waldner; p. 105 Gauche VIT S.A. (left) Soliver (right); p. 106: ECE; p. 107t: B. Chalmers; p. 107b: Super magnet; p. 109b: C. Oudet; p. 114: J. M. Schulz; p. 116t: M. F. Ashby, D. R. H. Jones; p. 116b: B. Chalmers; p. 117: CQFD International; p. 118: Personal collection.

Out-of-text notebook

I: Dapper Museum; II: Small Palace; III: Châtillon sur Seine Museum; IV: Archeological museum of Ankara; V: Nancy school museum; VI: DR; VII: Bernard Gasté/Les Chevaliers Carnutes; VIII: NOAA Corps Collection; IX: Thierry Baudin, X: Thierry Baudin; XI: D. Blavette.

Bibliography

This list of works is very limited. For more information on specific points in the book, readers can contact the author by email (louisette.priester@wanadoo.fr).

All Materials

A. Guinier (1984), *The Structure of Matter: From the Blue Sky to Liquid Crystals,* Butterworth-Heinemann Ltd.

A. Guinier, R. Julien, *Solid State: From Superconductors to Superalloys* (Oxford University Press, 1989)

M.-F. Ashby, D.R.H. Jones (1998 - 1999), *Engineering Materials* - Volume 1: *An Introduction to their Properties and Applications; Volume 2*: *An Introduction to Microstructures Processing and Design,* Butterworth-Heinemann Ltd.

M.-F. Ashby (2004), *Materials Selection in Mechanical Design,* Butterworth-Heinemann Ltd.

Support software. selection of materials: Ansys Granta Selector - Smarter Materials Choices

Metals and metallic alloys

P. Knauth (1974), *The metalsmiths,*Time-Life books.

J.P. Mohen, *The Bronze Age in Europe (New Horizons)* (Thames & Hudson, 2000)

J. Bénard, J. Michel, A. Philibert (1973), *Metalurgia general*, Editorial Hispano Europea.

L. Priester, *Materials: History, Science and Perspectives,*
https://doi.org/10.1007/978-3-032-15754-6

G. Pitron (2021), *The Rare Metals War: the dark side of clean energy and digital technologies,* B. Jacobsohn (Translation), Scribe Publications.

Ceramics et Glasses

Ph. BOCH (2007), *Ceramic Materials: Processes, Properties, and Applications,* Wiley-ISTE Pub.

Polymers

R.J. Crawford (1998), *Plastics Engineering*, Elsevier Ltd.
J. L. Leblanc (2010), *Filled Polymers Science and Industrial Applications,* CFR Press
S. Kulkarny, V. Shaikh (2025), *Specialty Polymers and Materials Advances, Technologies, Applications, and Future Trends,* Apple Academic Press.

The manufacturer's authorised representative in the EU is Springer Nature Customer Service Centre GmbH, Europaplatz 3, 69115 Heidelberg, Germany. If you have any concerns regarding our products, please contact ProductSafety@springernature.com

Printed and bound by CPI Group (UK) Ltd, Croydon, CR0 4YY
12/07/2026
02164711-0001